Protein Formulation
and Delivery

DRUGS AND THE PHARMACEUTICAL SCIENCES

Executive Editor

James Swarbrick

AAI, Inc.
Wilmington, North Carolina

Advisory Board

Larry L. Augsburger
University of Maryland
Baltimore, Maryland

David E. Nichols
Purdue University
West Lafayette, Indiana

Douwe D. Breimer
Gorlaeus Laboratories
Leiden, The Netherlands

Stephen G. Schulman
University of Florida
Gainesville, Florida

Trevor M. Jones
The Association of the
British Pharmaceutical Industry
London, United Kingdom

Jerome P. Skelly
Copley Pharmaceutical, Inc.
Canton, Massachusetts

Hans E. Junginger
Leiden/Amsterdam Center
for Drug Research
Leiden, The Netherlands

Felix Theeuwes
Alza Corporation
Palo Alto, California

Vincent H. L. Lee
University of Southern California
Los Angeles, California

Geoffrey T. Tucker
University of Sheffield
Royal Hallamshire Hospital
Sheffield, United Kingdom

Peter G. Welling
Institut de Recherche Jouveinal
Fresnes, France

DRUGS AND THE PHARMACEUTICAL SCIENCES

A Series of Textbooks and Monographs

1. Pharmacokinetics, *Milo Gibaldi and Donald Perrier*
2. Good Manufacturing Practices for Pharmaceuticals: A Plan for Total Quality Control, *Sidney H. Willig, Murray M. Tuckerman, and William S. Hitchings IV*
3. Microencapsulation, *edited by J. R. Nixon*
4. Drug Metabolism: Chemical and Biochemical Aspects, *Bernard Testa and Peter Jenner*
5. New Drugs: Discovery and Development, *edited by Alan A. Rubin*
6. Sustained and Controlled Release Drug Delivery Systems, *edited by Joseph R. Robinson*
7. Modern Pharmaceutics, *edited by Gilbert S. Banker and Christopher T. Rhodes*
8. Prescription Drugs in Short Supply: Case Histories, *Michael A. Schwartz*
9. Activated Charcoal: Antidotal and Other Medical Uses, *David O. Cooney*
10. Concepts in Drug Metabolism (in two parts), *edited by Peter Jenner and Bernard Testa*
11. Pharmaceutical Analysis: Modern Methods (in two parts), *edited by James W. Munson*
12. Techniques of Solubilization of Drugs, *edited by Samuel H. Yalkowsky*
13. Orphan Drugs, *edited by Fred E. Karch*
14. Novel Drug Delivery Systems: Fundamentals, Developmental Concepts, Biomedical Assessments, *Yie W. Chien*
15. Pharmacokinetics: Second Edition, Revised and Expanded, *Milo Gibaldi and Donald Perrier*
16. Good Manufacturing Practices for Pharmaceuticals: A Plan for Total Quality Control, Second Edition, Revised and Expanded, *Sidney H. Willig, Murray M. Tuckerman, and William S. Hitchings IV*
17. Formulation of Veterinary Dosage Forms, *edited by Jack Blodinger*
18. Dermatological Formulations: Percutaneous Absorption, *Brian W. Barry*
19. The Clinical Research Process in the Pharmaceutical Industry, *edited by Gary M. Matoren*
20. Microencapsulation and Related Drug Processes, *Patrick B. Deasy*
21. Drugs and Nutrients: The Interactive Effects, *edited by Daphne A. Roe and T. Colin Campbell*
22. Biotechnology of Industrial Antibiotics, *Erick J. Vandamme*
23. Pharmaceutical Process Validation, *edited by Bernard T. Loftus and Robert A. Nash*

Protein Formulation and Delivery

edited by

Eugene J. McNally
Boehringer Ingelheim Pharmaceuticals, Inc.
Ridgefield, Connecticut

MARCEL DEKKER, INC.   NEW YORK · BASEL

ISBN: 0-8247-7883-9 10180 4640

This book is printed on acid-free paper.

Headquarters
Marcel Dekker, Inc.
270 Madison Avenue, New York, NY 10016
tel: 212-696-9000; fax: 212-685-4540

Eastern Hemisphere Distribution
Marcel Dekker AG
Hutgasse 4, Postfach 812, CH-4001 Basel, Switzerland
tel: 41-61-261-8482; fax: 41-61-261-8896

World Wide Web
http://www.dekker.com

The publisher offers discounts on this book when ordered in bulk quantities. For more information, write to Special Sales/Professional Marketing at the headquarters address above.

Current printing (last digit):
10 9 8 7 6 5 4 3 2 1

Preface

This book is written to assist pharmaceutical scientists in the development of stable protein formulations starting early in the product development process, at which time only small amounts of drug substance are available and analytical methods are in the beginning stages of development. Books containing case studies of protein formulation and stability have been published over the last several years. These books summarize a wealth of information on the stability profile of different molecules that are reported in the literature only after the drug receives product approval. Little to no insight is provided on the approach that was used to identify the stable formulation nor is there any discussion of the formulation approaches that have failed. In contrast, this book directs scientists new to the field of protein formulation to appropriate starting points for their development efforts.

As a prerequisite to discussions of formulation, we start with an in-depth review of the mechanisms and associated causes of protein instability likely to be encountered during formulation development, including a discussion of accelerated stability testing and its limitations in identifying stable formulations. This is followed by a detailed discussion of analytical methods commonly used in stability assessment and formulation development, including a discussion of the importance of demonstrating the stability-indicating nature of an assay. Because many protein pharmaceuticals possess complex structure and functions, the analytical methods used to characterize the active ingredient and any unknown potentially harmful impurities are not always sufficiently sensitive to detect subtle changes in the protein at the molecular level. This limitation has led to stringent requirements regarding control of the manufacturing process, to ensure that the cell culture and purification processes produce a consistent product with an acceptable shelf-life. To this end, a brief description of the drug substance manufacturing process is given.

The next several chapters cover preformulation and development of traditional solution and lyophilized formulations often intended for intravenous administration. Specifically, we treat the issues of determining the reactivity of a molecule and the selection of formulation conditions, excipients, and container closure systems to minimize degradation processes and maximize shelf stability. Paramount to formulation selection is the development of the drug product manufacturing process, which traditionally involves aseptic processing, and potentially the development of a freeze-drying cycle. These manufacturing processes and the associated testing for evaluating their suitability for producing a stable product are also reviewed.

The last two chapters cover the special considerations involved in the development of nontraditional formulations directed toward alternative routes of drug delivery and controlled release dosage forms. These alternative routes of delivery are becoming increasingly important as biotechnology companies focus on the treatment of chronic indications for which frequent intravenous administration is not acceptable. Each of these chapters addresses the unique considerations specific to the route of delivery and the factors that need to be evaluated before a particular dosage form is chosen. For example, the physical and chemical characeristics that a protein must possess to be incorporated into a microsphere delivery system without affecting the activity of the molecule are discussed.

Overall, we provide a step-by-step approach to the development of solution and lyophilized protein formulations. In addition, we examine the additional considerations specific to more exotic formulations aimed at the nontraditional routes of delivery.

Eugene J. McNally

Contents

Contributors

Paul M. Bummer, Ph.D.   Department of Pharmaceutical Sciences, College of Pharmacy, University of Kentucky, Lexington, Kentucky

Michael L. Cappola, P.D.   Department of Pharmaceutics, Boehringer Ingelheim Pharmaceuticals, Inc., Ridgefield, Connecticut

Andrew R. Clark, Ph.D.   Department of Scientific Affairs, Inhale Therapeutic Systems, San Carlos, California

Helmut Hoffmann, Ph.D.   Department of Biopharmaceutical Manufacture, Boehringer Ingelheim Pharma KG, Biberach an der Riss, Germany

OluFunmi L. Johnson, Ph.D.   Formulation Development, Alkermes, Inc., Cambridge, Massachusetts

Sandy Koppenol, Ph.D.   Department of Bioengineering, University of Washington, Seattle, Washington

Christopher E. Lockwood, Ph.D.   Department of Pharmaceutics, Boehringer Ingelheim Pharmaceuticals, Inc., Ridgefield, Connecticut

Paul McGoff, M.S.   Department of Quality Control, Alkermes, Inc., Cambridge, Massachusetts

Eugene J. McNally, Ph.D.   Department of Pharmaceutics, Boehringer Ingelheim Pharmaceuticals, Inc., Ridgefield, Connecticut

David S. Scher, M.S.   Department of Formulation Development, Alkermes, Inc., Cambridge, Massachusetts

Steven J. Shire, Ph.D.   Department of Pharmaceutical Research and Development, Genentech, Inc., South San Francisco, California

James Wright, Ph.D.   Department of Development, Alkermes, Inc., Cambridge, Massachusetts

1
Overview of Protein Formulation and Delivery

James Wright
Alkermes, Inc.
Cambridge, Massachusetts

I. PROTEIN AND PEPTIDE REACTIONS

The experience of developing a number of protein and peptide formulations has taught me to look for three principal degradation pathways in formulations of proteins: deamidation, oxidation, and aggregation. In Chap. 2, Bummer and Koppenol cover these three pathways in depth. The problems associated with these protein degradation processes are extensively discussed in this monograph for good reason: they are the constant unwanted companions of development scientists in the protein delivery field.

The deamidation reaction is primarily the result of the hydrolysis of asparagine to aspartic acid. Thus, the reaction is most likely to proceed in an aqueous environment. If the protein is formulated in the solid state, the reaction rate is minimized. As discussed in Chap. 2, this reaction has been extensively studied, and the effects of pH, temperature, and buffers are well known and well documented. This extensive literature gives development scientists the opportunity to systematically optimize solution stability with respect to deamidation. In addition, the reactivity of asparagine is largely determined by neighboring residues, and thus the reactivity of the protein or peptide toward deamidation can be predicted.

Oxidation is in many ways a more complex process than deamidation in that the reactants and catalysts of the reaction are numerous and complex. There are several matters of significant concern; these include the ability to cause oxidation of specific residues through metal ion catalysis, by vapor phase hydrogen peroxide, photochemically, or through organic solvent, or even to diminish long-term stability in the presence of pharmaceutical excipients. Thus, each unit operation in the production and storage of a protein product is a potential source of oxidation.

The conformational stability of the protein is of fundamental concern in formulating proteins. Problems with conformational stability are often expressed as soluble aggregates and/or insoluble particulates (large aggregates). Aggregated proteins are a significant problem in that they are associated with decreased bioactivity and increased immunogenicity. The potential to produce aggregated forms is often enhanced by exposure of the protein to shear, liquid–air, liquid–solid, and even liquid–liquid interfaces. This means that there is potential to denature and aggregate a protein in almost any unit operation in the downstream processing of a protein, including the formulation of it. However, a protein can often be stabilized against aggregation by optimizing pH, temperature, and ionic strength, and through the use of surfactants.

II. PREFORMULATION AND ANALYTICAL DEVELOPMENT

The development of stability-indicating analytical methods for the drug product usually starts with the preformulation studies. During the preformulation studies, methods are selected and the first evaluation of the protein on short-term stability is conducted. Since one of the most significant purposes of the pre-formulation work is to define the processes that destroy protein integrity and activity, it is critical that the analytical methods chosen for the task indeed be stability indicating. The three basic protein degradation processes (deamidation, oxidation, and aggregation) must be followed by accurate, precise, and simple analytical methods.

Chapters 3 and 4 review analytical test methodology and studies preformulation for proteins and peptides. Deamidation of a protein or a peptide is a common and important chemical decomposition reaction. The deamidation of a protein can be followed by two simple techniques: isoelectric focusing and ion exchange chromatography. Each of these analytical techniques allows the development scientist to follow the disappearance of the starting isoforms and the generation of reaction products. Often, these reaction products (the more deamidated state of the molecule) are easy to isolate and test in bioactivity assays and in pharmacokinetic studies. These activity studies are important steps to take when there is concern that the molecule is prone to deamidation, since the practical consequences of deamidation can be judged scientifically.

Oxidation is often determined using reversed-phase HPLC or by peptide analysis. Aggregation is usually determined by size exclusion chromatography (SEC) or by sodium dodecyl sulfate–polyacrylamide gel electrophoresis (SDS-PAGE). These two techniques give different answers because they address different questions. SDS-PAGE detects and quantifies irreversible aggregates, while SEC will detect noncovalent aggregates as well. The distinction is critical, since the stability of the dosage form may be judged by either or both methods.

During the preformulation and analytical development phase, there is opportunity to determine the correlation between the different analytical techniques and the activity assay. There will be an activity assay that judges the suitability and stability of the dosage form developed. It is imperative to know early in the development process which analytical techniques predict and which degradation processes correlate with losses in activity.

III. SOLUTION FORMULATION STABILITY

In Chap. 5, McGoff and Scher review the issues of protein solubility, aggregation, and adsorption. The authors provide a systematic framework for evaluating the physical stability of proteins. The problems associated with physical instability are serious in that they lead to aggregation, adsorption to surfaces, and loss of biological activity. Thus, formulation studies to determine optimal stabilization conditions should include variables such as temperature, solution pH, buffer ion, salt concentration, protein concentration, and the effect of surfactants. In addition, the protein should be characterized and stabilized against adsorption and/or denaturation at interfaces.

The effect of concentration of the protein on stability of the formulation is also a critical concern. Proteins undergo concentration-dependent aggregation and adsorption, and thus the effect of protein concentration on physical stability must be accounted for. The testing of the adsorption potential of a protein with the range of materials that might be used in packaging or processing is an important part of the formulation effort.

Freeze/thaw stability is also a critical issue for both the formulation and the storage of bulk protein. The development of a liquid formulation for aerosol delivery requires the consideration of additional routes of degradation of the formulated protein. In Chap. 7 on pulmonary delivery, Clark and Shire point out that generating respirable droplets for pulmonary delivery also exposes the protein to the air–water interface and to shear. Thus the formulation scientist must work with methods to reduce this exposure through the use of surfactants.

IV. SOLID STATE FORMULATION STABILITY

The pharmaceutical solid state product of a protein or peptide is usually produced by freeze-drying. A freeze-dried product is produced by a process that,
if done correctly, is friendly to protein structure and activity and can prolong
the shelf life of the product. The degradation processes such as deamidation,
oxidation, aggregation, and hydrolysis are all significantly reduced in a dried
solid state protein product. The reason that proteins have enhanced stability
in the solid state is that molecular mobility is drastically reduced.

A protein is usually present as an amorphous solid in a lyophilized product, with a glass transition temperature associated with the material. In very
rough terms, at temperatures above the glass transition temperature there is
significant molecular mobility, while below the glass transition temperature,
reduced mobility and reaction rates are achieved. Thus, the stabilization of a
protein is dependent on the use of a freeze-drying process and formulation
that enhances protein conformation and reduces molecular mobility. Reduced
mobility and enhanced protein stability is achieved through temperature, moisture, and excipient control during and after the freeze-drying process. The
effect and optimal selection of process conditions and formulation components
on protein stabilization are outlined in Chap. 6 on freeze-drying. Chapters 7
and 8 discuss the effect of solid state protein stabilization on pulmonary and
controlled delivery.

V. PROTEIN DELIVERY

Some very interesting and challenging problems are associated with drug delivery of proteins. One of the first issues to be faced is the stability of the
protein in the body, inasmuch as most drug delivery systems for proteins are
implanted or injected by the intramuscular or subcutaneous routes. The body
provides an ideal environment for the degradation of a protein, and thus the
stability of the product after administration is a critical concern. The 37°C
temperature and neutral pH conditions of the human body encourage deamidation reactions. In addition, the higher temperature coupled with moisture will
promote molecular mobility and thus accelerate degradation reactions. In addition, the formulation and delivery system must be designed to minimize tissue
damage and irritation, since inflammation can also provide routes of protein
and peptide degradation. Thus, the formulation of proteins and peptides must
also account for response and reaction in biological tissues.

2
Chemical and Physical Considerations in Protein and Peptide Stability

Paul M. Bummer
University of Kentucky
Lexington, Kentucky

Sandy Koppenol
University of Washington
Seattle, Washington

I. DEAMIDATION

A. Introduction

The deamidation reactions of asparagine (Asn) and glutamine (Gln) side chains are among the most widely studied nonenzymatic covalent modifications to proteins and peptides (1–7). Considerable research efforts have been extended to elucidate the details of the deamidation reaction in both in vitro and in vivo systems, and a number of well-written, in-depth reviews are available (1–5,8,9). This work touches only on some of the highlights of the reaction and on the roles played by pH, temperature, buffer, and other formulation components. Possible deamidation-associated changes in the protein structure and state of aggregation also are examined. The emphasis is on asparagine deamidation, since glutamine is significantly less reactive.

B. REACTION MECHANISM

The primary reaction mechanism for the deamidation of Asn in water-accessible regions of peptides and proteins at neutral pH is shown in Fig. 1. For the present, discussion is confined to the intramolecular mechanism, uncomplicated by adjacent amino acids at other points in the primary sequence. The key step is the formation of a deprotonated amide nitrogen, which carries out the rate-determining nucleophilic attack on the side chain carbonyl, resulting in the formation of the five-membered ring succinimide intermediate. For such

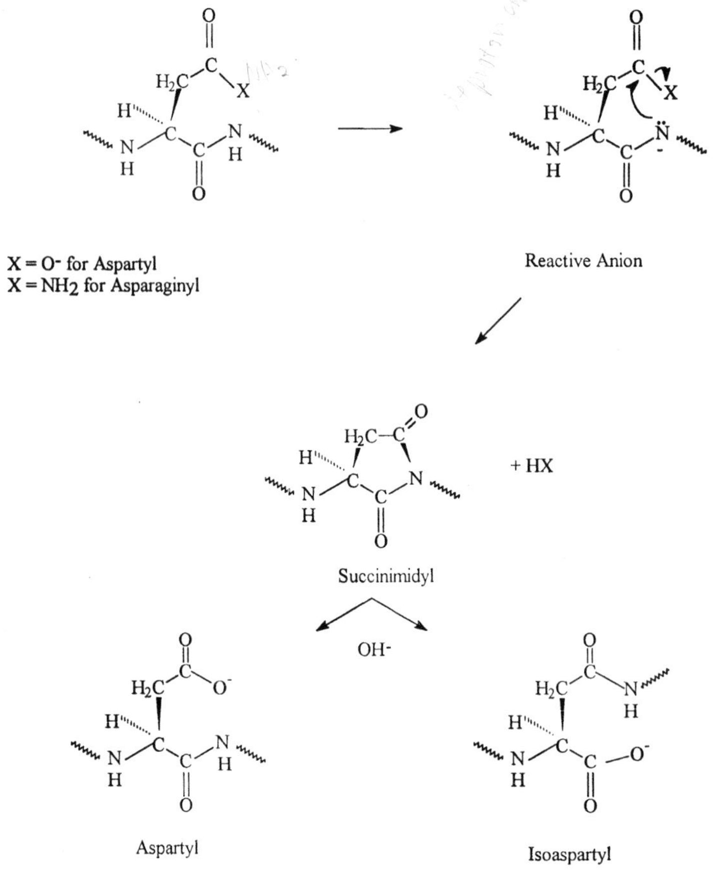

X = O⁻ for Aspartyl
X = NH₂ for Asparaginyl

Reactive Anion

Succinimidyl

+ HX

OH⁻

Aspartyl

Isoaspartyl

Fig. 1 Proposed reaction mechanism for deamidation. Note the formation of the succinimidyl intermediate.

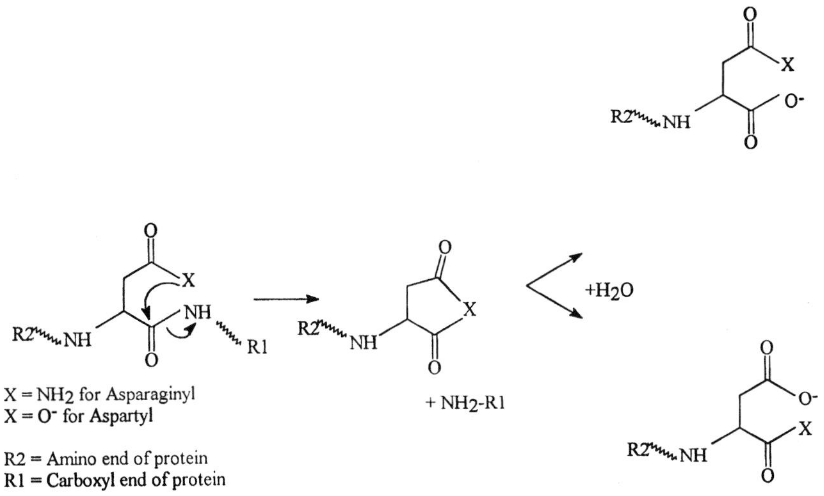

Fig. 2 Proposed reaction mechanism for main chain cleavage by aspartyl or asparaginyl residues.

a reaction, the leaving group must be easily protonated, and in this case it is responsible for the formation of ammonia (NH_3). The succinimide ring intermediate is subject to hydrolysis, resulting in either the corresponding aspartic acid or the isoaspartic acid (β-aspartate). Often, the ratio of the products is 3:1, isoaspartate to aspartate (10–12). The reaction also appears to be sensitive to racemization at the α carbon, resulting in mixtures of D- and L-isomers (10,13,14). No matter the final product, the rate of degradation of the parent peptide in aqueous media most often follows pseudo-first-order kinetics (15,16).

A number of other alternative reactions are possible. The most prevelant appears to be a nucleophilic attack of the Asn side chain amide nitrogen on the peptide carbonyl, resulting in main chain cleavage (10,15,17). This reaction (see Fig. 2) is slower than that of cyclic imide formation and is most frequently observed when Asn is followed by proline, a residue incapable of forming an ionized peptide bond nitrogen.

C. pH Dependence

Under conditions of strong acid (pH 1–2), deamidation by direct hydrolysis of the amide side chain becomes more favorable than formation of cyclic imide

(15,18). Under these extreme conditions, the reaction is often complicated by main chain cleavage. Deamidation by this mechanism is not likely to produce isoaspartate or significant racemization (15).

In the cyclic imide mechanism present under more moderate conditions, the effect of pH is the result of two opposing reactions: (a) deprotonation of the peptide bond nitrogen promoting the reaction and (b) protonation of the side chain leaving group, inhibiting the reaction. In deamidation reactions of short chain peptides uncomplicated by structural alterations or covalent dimerization (19), the pH–rate profiles exhibit the expected "V" shape with a minimum occurring in the pH range of 3–4 (15). The increase in rate on the alkaline side of the minimum does not strictly correlate with the increase in deprotonation of the amide nitrogen, indicating that the rate of reaction is not solely dependent on the degree of the peptide bond nitrogen deprotonation (15,18). The pH minimum in the deamidation reaction measured in vitro for proteins may (20) or may not (21) fall in the same range as that of simple peptides. Overall pH-dependent effects may be modified by additional, structure-dependent factors, such as dihedral angle flexibility, water accessibility, and proximity of neighboring amino acid side chains (see below).

D. Effect of Temperature

The temperature dependence of deamidation rate has been studied in a variety of simple peptides in solution (15,22,23). Small peptides are easily designed to avoid competing reactions, such as oxidation and main chain cleavage, and are thus useful to isolate attention directly on the deamidation rate. In solution, deamidation of small peptides tends to follow an Arrhenius relationship. Activation energies of the reaction do tend to show pH dependence, and a discontinuity in the Arrhenius plot is expected when the mechanism changes from direct hydrolysis (acid pH) to one of cyclic imide (neutral to alkaline pH).

The deamidation rate of proteins also shows temperature dependence (21,24,25) under neutral pH. The inactivation of proteins as a function of temperature is much more difficult to isolate solely to deamination for several reasons: competing reactions at other side chains, thermal instability of the structure of the protein, or main chain hydrolysis. For deamination reactions alone, rate acceleration in protein may be due to enhanced flexibility in the protein, allowing more rapid formation of the cyclic imide formation (26), or it may occur by catalysis by side chains brought into the vicinity of the deamidation site (5).

The availability of water appears to be an important determinant in temperature-associated effects. In studies of lyophilized formulation of Val-Tyr-

Pro-Asn-Gly-Ala, the deamination rate constant was observed to increase about an order of magnitude between 40 and 70°C (27). In contrast, in the solid state, the Arrhenius relationship was not observed. Further, the deamidation in the solid state showed a marked dependence upon the temperature when the peptide was lyophilized from a solution of pH 8, while little temperature dependence was observed when lyophilization proceeded from solutions at either pH 3.5 or 5. The authors related this temperature difference to the difference in mechanism that may occur as a function of pH. In direct hydrolysis at low pH, water is a necessary reactant. In the cyclic imide mechanism at neutral to alkaline pH, deamidation can occur without the availability of water.

E. Adjuvants and Excipients

The influence on deamidation by a variety of buffer ions and solvents has been examined. As pointed out by Cleland et al. (4) and reinforced by Tomizawa et al. (13), many of these additives are unlikely to be employed as pharmaceutical excipients for formulation but may be employed in protein isolation and purification procedures (28). Important clues to stabilization strategies can be gained from these studies. In all the following, it is fruitful to keep in mind the importance of the attack of the ionized peptide bond nitrogen on the side chain carbonyl and the hydrolysis of the cyclic imide (Fig. 1).

1. Buffers

Buffer catalysis appears to occur in some but not all peptides and proteins (5). Bicarbonate (15) and glycine (12) buffers appear to accelerate deamidation. On one hand, phosphate ion has been shown to catalyze deamidation, both in peptides and proteins (12,13,15,29–31), generally in the concentration range of 0–20 mM. On the other hand, Lura and Schrich (32) found no influence on the rate of deamidation of Val-Asn-Gly-Ala when buffer components (phosphate, carbonate, or imidazole) were varied from 0 to 50 mM. A general acid–base mechanism by which phosphate ion catalyzes deamidation was challenged in 1995 by Tomizawa et al. (13), who found that the deamidation rate of lysozyme at 100°C did not exhibit the expected linear relationship of deamidation rate on phosphate concentration. Brennan and Clarke (16) have suggested that phosphate ion, or other buffer ingredients, could act on the aqueous solvent to increase basicity of water molecules without forming free hydroxide ions. Experimental data directly supporting this speculation have not yet appeared in the literature.

2. Ionic Strength

The effects of ionic strength appear to be complicated and not open to easy generalizations. Buffer and ionic strength effects on deamidation are evident in proteins at neutral to alkaline pH (5). In selected peptides and proteins, the catalytic activity of phosphate has been shown to be reduced moderately in the presence of salts NaCl, LiCl and Tris HCl (12,13). Of these salts, NaCl showed the least protective activity against deamidation (13).

In the peptide Gly-Arg-Asn-Gly at pH 10, 37°C, the half-life $t_{1/2}$ of deamidation dropped from 60 hours to 20 hours when the ionic strength was increased from 0.1 to 1.2 (20). However, in the case of Val-Ser-Asn-Gly-Val at pH 8, 60°C, there was no observable difference in the $t_{1/2}$ of deamidation when solutions without salt were compared to those containing 1 M NaCl or LiCl (12). Interestingly, for lysozyme at pH 4 and 100°C, added salt showed a protective effect against deamidation, but only in the presence of phosphate ion (13).

In reviewing the data above, Brennan and Clarke (16) tentatively attributed the promotion of deamidation by elevated levels of ions to enhanced stabilization of the ionized peptide bond nitrogen, promoting attack on the side chain amide carbonyl. Other mechanisms would include disruption of tertiary structure in proteins that may have stabilized Asn residues, in some as-yet unknown fashion. That promotion of deamidation is observed in some cases of peptides and inhibition in others does suggest rather complex and competing effects. Clearly the stabilizing effects, when observed at all, are often at levels of salt too concentrated for most pharmaceutical formulations.

3. Solvents

The effect of various organic solvents on the rate of deamidation has not received much attention; it would be expected, however, that in the presence of a reduced dielectric medium, the peptide bond nitrogen would be less likely to ionize. Since the anionic peptide bond nitrogen is necessary in the formation of the cyclic imide, a low dielectric medium would retard the progress of the reaction and be reflected in the free energy difference for ionization of the peptide bond nitrogen (16). Following this hypothesis, Brennan and Clarke (33) analyzed succinimide formation of the peptide Val-Tyr-Pro-Asn-Gly-Ala [the same peptide employed by Patel and Borchardt (15) in studies of pH effects in aqueous solution] as a function of organic cosolvent (ethanol, glycerol, and dioxane) at constant pH and ionic strength. The lower dielectric constant media resulted in significantly lower rate of deamidation, in agreement with the hypothesis. It was argued that the similar rates of deamidation for different cosolvent systems of the same effective dielectric constant indicated that changes in viscosity and water content of the medium did not play a significant role.

The effect of organic cosolvents on deamidation in proteins is even less well characterized than that of peptides. Trifluoroethanol (TFE) inhibits deamidation of lysozyme at pH 6 and 100°C (13), and of the dipeptide Asn-Gly, but does not inhibit the deamidation of free amino acids. The mechanism of protection is not clear; direct interaction of the TFE with the peptide bond was postulated, but not demonstrated. An alternative hypothesis is that TFE induces greater structural rigidity in the protein, producing a structure somewhat resistant to the formation of the cyclic imide intermediate. Other, pharmaceutically acceptable alcohols, ethanol and glycerin, did not exhibit the same protective effects as TFE on lysozyme.

Of course, in dosage form design, organic solvents such as TFE do not make for attractive pharmaceutical adjuvants. The effects of low dielectric may still supply a rational for the solubilization of peptides in aqueous surfactant systems, where the hydrophobic region of the micell or a liposome could potentially enhance stabilization of the Asn residues from deamidation.

As pointed out by Brennan and Clarke (16), the results of experiments in organic solvents can have implications on prediction of points of deamidation in proteins as well. For Asn residues near the surface of the protein, where the dielectric constant is expected to approach that of water (a value of 78 at 25°C; Ref. 34), deamidation rate would be expected to be high. For Asn residues buried in more hydrophobic regions of the protein, where polarities are thought to be more in line with that of ethanol or dioxane (35), reaction rates would be expected to be considerably slower.

F.   Peptide and Protein Structure

The ability to identify which Asn or Gln residues in a therapeutic protein or peptide vulnerable to deamidation would have great practical application in preformulation and formulation studies. The effects of various levels of structure—primary, secondary, and tertiary—are believed to be complex and varied. At present, only primary structure effects have been characterized in a systematic manner.

1.   Primary Sequence

The primary sequence of amino acids in a peptide or protein is often the first piece of structural data presented to the formulation scientist. Considerable effort has been extended to illucidate the influence of flanking amino acids on the rates of deamidation of Asn and Gln residues. The potential effects of flanking amino acids are best elucidated in simple peptides, uncomplicated by side reactions or secondary and tertiary structure effects.

a. Effect of Amino Acids Preceding Asn or Gln

In an extended series of early studies, Robinson and coworkers (22) examined the influence of primary sequence on the deamidation of Asn or Gln in the middle of a variety of pentapeptides. Mild physiologic conditions (pH 7.5 phosphate buffer at 37°C) were employed. A few general rules can be extracted from this work:

1. In practically every combination tested, Gln residues were less prone to deamidation than Asn. For the two residues placed in the middle of otherwise identical host peptides, the half-life of the reactions differed by a factor ranging from two- to three-fold.
2. In peptides Gly-X-Asn-Ala-Gly, steric hinderance by un-ionized X side chains inhibits deamidation. The rank order of deamidation rate found was Gly > Ala > Val > Leu > Ile with the $t_{1/2}$ ranging from 87 to 507 days. It remains unclear why bulky residues inhibit the reaction, but reduced flexibility of the sequence may be a factor. A similar effect was noted when Gln replaced Asn. In this later case, $t_{1/2}$ ranged from 418 to 3278 days, in accordance with the diminished reactivity of glutamine.
3. For the same host peptide, when X side chain was charged, the deamidation rate of Asn followed the rank order of Asp > Glu > Lys > Arg.

b. Effect of Amino Acids Following Asn or Gln

Early experiments on dipeptides under extreme conditions indicated a particular vulnerability of the Asn-Gly sequence to deamidation (36). More recent studies of ACTH-like sequence hexapeptide Val-Tyr-Pro-Asn-Gly-Ala under physiologic conditions (37) has verified that deamidation is extremely rapid ($t_{1/2}$ of 1.4 days at 37°C). The formation of the succinimide intermediate is thought to be the basis for sequence dependence (10) of deamidation. It is generally believed that bulky residues following Asn may inhibit sterically the formation of the succinimide intermediate in the deamidation reaction. Replacement of the glycyl residue with the more bulky leucyl or prolyl residues resulted in a 33- to 50-fold (respectively) decrease in the rate of deamidation (10). Owing to the highly flexible nature of the dipeptide, the deamidation rate observed in Asn-Gly is thought to represent a lower limit.

Steric hinderance of the cyclic imide formation is not the only possible genesis of sequence-dependent deamidation. The resistance to cyclic imide formation in the presence of a carboxyl-flanking proline peptide may be related to the inability of the prolyl amide nitrogen to attack the Asn side chain (10). For other residues, electron-inducing effects of the side chain of the following residue may inhibit deprotonation of the peptide bond nitrogen, inhibiting deamidation (15,38).

In more recent studies, deamidation of Val-Tyr-X-Asn-Y-Ala, a peptide sequence derived from ACTH, the adrenocorticotropic hormone, was examined with different residues in both flanking positions (39). When X is histidine (and Y is glycine), no acceleration of deamidation was found relative to a peptide where X is proline. Placing a His following the Asn was found to result in similar rates of deamidation when X was phenylalanine, leucine, or valine. The rate when X was histidine was slower than that of alanine, cysteine, serine, or glycine. These results indicate that histidine does not have unique properties in facilitating succinimide formation. Of interest was the observation that histidine on the carboxyl side of the asparagine did seem to accelerate main chain cleavage products.

Some of the general rules for peptides may also show higher levels of dependence on primary sequence. Tyler-Cross and Schirch (12) studied the influence of different amino acids on the adjacent amino end of the pentapeptide Val-X-Asn-Ser-Val at pH 7.3. For X = His, Ser, Ala, Arg, and Leu, deamidation rates were essentially constant and approximately seven times slower than the Val-Ser-Asn-Gly-Val standard peptide. Of special interest to the investigators was the observation of no difference in deamidation rates between those amino acids with and without β-branching (such as valine for glycine). This is in direct contrast to the findings of Robinson and Rudd (22) of 10-fold differences in deamidation for valine substitution for glycine in Gly-X-Asn-Ala-Gly, shown earlier. Under the mild alkaline conditions of Patel and Borchardt (15), Val-Tyr-X-Asn-Y-Ala, no difference in the deamidation rate constants was observed when proline was substituted for glycine in the X position.

2. Secondary and Tertiary Structure

X-ray or NMR data can provide a detailed map of the three-dimensional structure of the protein or peptide. The role of secondary and tertiary structures in intramolecular deamidation of proteins has been discussed by Chazin and Kossiakoff (40). It is beyond the intent of this work to present a comprehensive review of the details of deamidation reactions in specific proteins. Excellent reviews of a variety of specific proteins exist (see, e.g., Ref. 6). For the most part, detailed mechanisms relating the secondary and tertiary structure of proteins to enhancement of rates of deamidation are not yet available.

Clear differences in deamidation rates of some proteins are evident when native and denatured states are compared (13,41). Denaturation is thought to enhance main chain flexibility and water accessibility (40). Sufficient conformational flexibility is required for the Asn peptide to assume the dihedral angles of $\Phi = -120°C$ and $\Psi = +120°C$ necessary for succinimide formation. Inasmuch as such angles tend to be energetically unfavorable (42) in native proteins, it may be expected that Asn residues in the midst of rigid secondary structures, such as helices, may be resistant to deamidation.

The direct influence of secondary structure on deamidation may be best understood in terms of hydrogen-bonding patterns that give rise to defined structures. The α-helix is characterized by the hydrogen-bonding of the main chain carbonyl oxygen of each residue to the backbone NH of the fourth residue along the chain. The resulting bond is close to the optimal geometry, and therefore maximal energy, for such an interaction (42). Hydrogen bonds in β-sheets are not of fixed periodicity as in the helix, but can exhibit comparable bond energies. Citing structural data for trypsin (43), Chazin and Kossiakoff (40) argue that strong main chain hydrogen bonding of the peptide nitrogen following Asn is an important factor in modulating deamidation. Since formation of the succinimide intermediate requires the peptide nitrogen to be free to attack the side chain carbonyl, participation in a strong hydrogen bond by that nitrogen would inhibit the reaction. X-ray crystallography or NMR data may be helpful in identifying Asn residues in native structures likely to be protected by such a mechanism (40). Perhaps studies modeled along the lines of guest–host relationships would be helpful in elucidating further the influence of secondary structure on deamidation (44).

G. Effects of Deamidation on Structure

In 1994, in an extensive and detailed series of studies, Darrington and Anderson showed that deamidation strongly influences the noncovalent self-association (45) and covalent dimer formation (45,46) of human insulin. The noncovalent dimer formation of triosephosphatase (47) is inhibited by deamidation, probably by charge repulsion arising from the resulting additional anionic charges present in the hydrophobic faces of the monomers.

Deamidation in concentrated solutions of food proteins tends to show increased viscosity, possibly due to enhanced charge interactions between formerly uncharged portions of the protein molecule (41). The isoelectric point of the deamidated molecule is shifted toward lower values, possibly resulting in modified potential for adsorption to solid surfaces (48). Foamability of protein solutions subject to deamidation is greatly enhanced, probably because of partial unfolding (41).

Deamidation can destabilize a protein, making thermal (47) or chemical (13) denaturation more likely. Folding patterns may be influenced (49,50), and changes in secondary structure can result (47). Other proteins appear to be resistant to structure alterations secondary to deamidation (51–56).

II. OXIDATION

A. Introduction

Oxidation has been identified as one of the major degradation pathways in proteins and peptides and can occur during all steps of processing, from protein isolation to purification and storage (57,58). An extensive listing of proteins undergoing oxidation, the primary amino acids involved, and the biological activity of the products has been provided (4). A change in the biological activity of a therapeutic agent potentially can arise from an altered enzymatic activity, inhibited receptor binding properties, enhanced antigenicity, or increased sensitivity to in vivo proteases. In some instances, biological activity is completely or partially lost upon oxidation, while in other instances, no effect on bioactivity is observed. The molecular mechanism of altered bioactivity often comes about either by oxidation of a critical residue at or near the enzyme active site or receptor binding site, or by a dramatic change in the structure of the protein upon oxidation. At present, no general rules are evident to predict with certainty all the effects of oxidation on the biological activity of a particular protein.

The chemistry of autoxidation (i.e., oxidation not enzyme- or radiation-catalyzed) in nonprotein drug molecules has been reviewed (59,60). There are three main steps that make up any free radical chain reaction oxidation mechanism, namely: initiation, propagation, and termination. In the initiation step, free radical generation is catalyzed by transition metal ions, light energy, or thermal energy. Once initiated, oxidation reactions propagate by chain reactions of organic substances with reactive oxygen species such as singlet oxygen, hydroxyl, and peroxyl radicals. The propagation steps are either hydrogen atom abstraction or addition to olefin. In the termination step, free radicals, both alkyl and reactive oxygen, are consumed without producing further radicals among the products. For the purposes of pharmaceuticals, it is important to emphasize the role of both trace metal ions and dissolved oxygen in accelerating oxidation (57,61).

B. Oxidation in Pharmaceutical Proteins and Peptides

In living systems, a variety of well-characterized reactive oxygens are produced (62,63). In pharmaceutical formulations, identifying a single oxidation initiator is often difficult, since a variety of initiation possibilities exist, such as photochemical (4,64,65), metal-ion-catalyzed (66,67), high energy γ-radiation (66), and organic additives (68,69). Even something as seemingly simple as sonication appears to promote the generation of reactive oxygen species (70,71). It has been convincingly shown that the extent of protein oxidation, and subsequent loss of biological activity, exhibits strong dependence upon the oxidation system employed (65,72–74).

In pharmaceutical proteins, transition metal ion catalysis of oxidation has received the lion's share of attention (73,75,76), while much less attention has been devoted to light energy and thermal energy (64,65).

1. Metal Ion Catalysis of Oxidation

Because of their importance in biological systems, a variety of metal-ion-catalyzed oxidation systems have been identified and cataloged (66). Since the metal-ion-catalyzed systems tend to be amenable to laboratory manipulations, they have been employed in stability studies (73,75,76). More importantly, trace levels of metal ions known to initiate oxidation are often present as contaminants in pharmaceutical systems (57), making an understanding of metal ion catalysis highly relevant to the job of formulation stabilization.

Iron(II) and copper(II) salts, in the presence of molecular oxygen and water, will slowly oxidize to form $O_2^{\cdot-}$ (superoxide radical) by Eq. 1.

$$Fe(II) + O_2 \Leftrightarrow Fe(III) + O_2^{\cdot-} \tag{1}$$

The superoxide radical is not stable at neutral pH and undergoes dismutation to form hydrogen peroxide by Eq. 2.

$$2OO_2^{\cdot-} + 2H^+ \Rightarrow H_2O_2 + O_2 \tag{2}$$

Hydrogen peroxide reacts further to produce hydroxyl radicals (OH·) by Eq. 3.

$$H_2O_2 + Fe(II) \Rightarrow Fe(III) + OH^- + OH\cdot \tag{3}$$

Hydroxyl radicals are capable of abstracting hydrogen atoms with bond energies less than 89 kcal/mol (77–79) producing organic radicals by Eq. 4.

$$OH\cdot + RH \Rightarrow H_2O + ROO\cdot \tag{4}$$

The organic radical (ROO·) is capable of entering into a whole variety of chain reaction propagation and termination reactions (75). Overall, Eqs. 1–4 show the production of four different reactive oxygen species, each able to oxidize pharmaceutical proteins. In solutions of free amino acids, oxidation by OH· shows a strong dependence on bicarbonate ion (66,80). It has been suggested that the bicarbonate ion may be required to interact with the amino acid and Fe(II) to form a hybrid complex. Bicarbonate appears not to be a necessary reactant of protein or peptide oxidation.

2. Site-Specific Metal-Ion-catalyzed Oxidation

Radiolysis studies have shown that all amino acid side chains are vulnerable to oxidation by reactive oxygen species. The same oxygen radicals, when produced by metals (Esq. 1–4), tend to attack preferentially only a few amino acid residues, most notably His, Met, Cys, and Trp. In addition, metal-ion-catalyzed oxidation of proteins can show relative insensitivity to inhibition by free radical scavenger agents (75,76). These observations have led to the hypothesis that metal-ion-catalyzed oxidation reactions are "caged" processes in which amino acid residues in the immediate vicinity of a metal ion binding site are specific targets of the locally produced reactive oxygen. Schoneich and Borchardt have discussed the following reaction (76):

$$D\text{--}Fe(II) + O_2 + H^+ \Rightarrow D\text{--}Fe(III)\text{--}OOH \tag{5}$$

where D is some binding ligand, such as a buffer species, peptide, or protein. By this mechanism, any amino acid residues capable of forming a metal ion binding site are potential sources of reactive oxygen species. Since reaction of the oxygen radical usually occurs in the immediate region of its production before escape into the bulk solution by diffusion, free radical scavengers are unlikely to be effective formulation protective agents (76). The terminal hydroxyl group of serine, the free carboxyl group of aspartic and glutamic acids, the imidazole ring of histidine, and the free amino or free carboxyl groups of N-terminal or C-terminal (respectively) residues have all been suggested to participate in binding metal ions to proteins (80). Further, since a metal ion binding site may be formed by appropriate residues upon folding of the protein molecule, these amino acids need not be adjacent to each other in the primary sequence.

3. Oxidation by Hydrogen Peroxide Addition

Addition of hydrogen peroxide has been employed as a means to study oxidation of proteins (72,81,82), the advantage being that the concentration and identity of the initiating oxidant is known. In some instances, hydrogen peroxide has been shown to be an oxidant specific for methionine (83) while in other instances, oxidation of cysteine and tryptophan residues also occurs (84). Hydrogen peroxide is thought to oxidize only residues easily accessible on the surface of the folded protein, but more recent evidence suggests oxidation of both surface and buried residues (83). It has been proposed that t-butyl hydroperoxide may be a highly specific oxidizer of surface-localized methionine residues (83).

C. Specific Amino Acid Side Chains

1. Methionine

Methionine has been identified as one of the most easily oxidized amino acids in proteins, and oxidation of this residue has received considerable attention. Oxidation deprives methionine of its ability to act as a methyl donor, which will influence the bioactivity of proteins dependent on that function (85). The reaction product of methionine oxidation is the corresponding sulfoxide and, under more strenuous oxidation conditions, the sulfone (Fig. 3). These are not the only possible reaction products, but they are usually the first to appear.

Not surprisingly, mechanisms of oxidation of methionine appear to be highly dependent on the reactive oxygen species under consideration (65). Peroxide (86), peroxyl radicals (87), singlet oxygen (86), and hydroxyl radical (75) have all been shown to oxidize methionine residues to sulfoxides and other products. The identity of major oxidizing species present in these solutions remains a matter of controversy (88, 89).

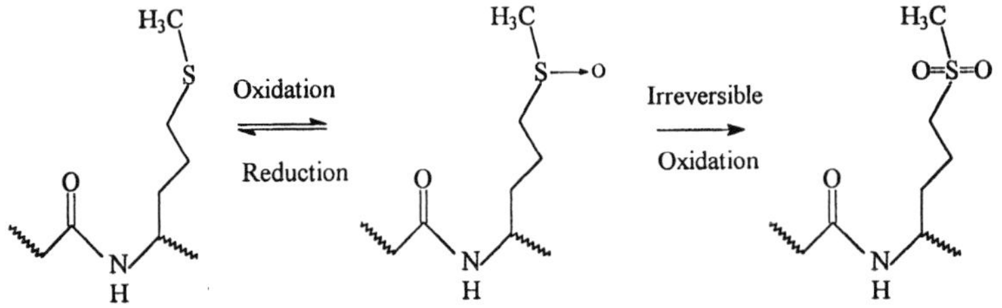

Fig. 3 Oxidation of methionine first to the sulfoxide and then to sulfone derivatives.

The reaction mechanisms for proteins in pharmaceutical systems are incomplete because not all products and intermediates are known. An excellent example of oxidation induced by a variety of pro-oxidants in a single recombinant protein and the experimental difficulties encountered is presented next.

a. Oxidation of Methionine in Recombinant Human Relaxin

Photocatalyzed oxidation. A series of papers spanning the 1990s studied methionine oxidation initiated by light (65), hydrogen peroxide (72), and ascorbic acid–Cu(II) (73) in recombinant human relaxin. Upon exposure to light of an intensity of 3600 candles for 5–17 days, both methionine residues, Met-B4 and Met-B25, located on the surface region of the B-chain, were oxidized to the sulfoxide derivative (65). The identity of the reactive oxygen species formed upon exposure to light was not reported, but peptide mapping results suggest a wide variety of reaction products.

Hydrogen Peroxide. In the presence of added hydrogen peroxide, the methionines (Met-B4 and Met-B25) were the only residues of relaxin to be oxidized (72). Three products were isolated, the monosulfoxide at either methionine and the corresponding disulfoxide. The reaction rate was independent of pH (range 3–8), ionic strength (0.007–0.21 M NaCl), or buffer species (lactate, acetate, Tris). Interestingly, the rate of reaction of the two methionine groups differed, with oxidation at Met-B25 more rapid than that at Met-B4. The oxidation rate of Met-B25 was equivalent to that observed for free methionine and for methionine in a model peptide of the relaxin B-chain (B23–B27). The reduced rate of oxidation at the solvent-exposed residue Met-B4 relative to Met-B25 suggests that accessibility of the residues to H_2O_2 may play a role in the reaction.

Pro-oxidant System Ascorbic Acid–Cu(II). Contrary to the results observed in the presence of hydrogen peroxide, in the presence of the pro-oxidant system ascorbic acid–Cu(II), a pH-dependent precipitation of relaxin was observed (73). Approximately 80% of protein was lost from solution within 25 minutes at pH 7–8. Chromatographic results indicated that the aggregate formed was not held together by covalent forces. In a second significant contrast to the study above (72), in the presence of ascorbic acid–Cu(II), investigators observed oxidation of histadine, methionine, and lysine (73). One final important difference is that Met-B4 was oxidized preferentially over Met-B25. All these differences are consistent with the conclusion that the oxidant system employed for in vitro studies can have a major impact on the results. Clearly, the issue of identifying the radical species responsible for oxidation of methionine, or any other residue, is of primary importance in setting down complete reaction mechanisms.

b.   Methionine Oxidation Studies with Model Peptides

As has been pointed out, development of a molecular-level understanding of oxidation in protein drug delivery systems has been hampered by a lack of characterization of the reaction mechanism and the products. Trailblazing work of Li, Schoneich, and Borchardt, and their colleagues (75,76,90) has begun to address the much-needed mechanistic description of the effects of pH and primary sequence on oxidation pathways of methionine in simple model peptides. These authors have primarily employed the metal-ion-catalyzed pro-oxidant system and a series of simple methionine-containing peptides. Considerable efforts have been expended with specific radical scavengers to identify the reactive oxygen species responsible for oxidation.

c.   Buffers and pH

Using the pro-oxidant systems of DTT/Fe (III) to generate reactive oxygen species, oxidation of methionine in Gly-Gly-Met, Gly-Met-Gly, and Met-Gly-Gly was studied as a function of pH. The degradation of all peptides followed first-order kinetics, while mass balance comparisons showed that sulfoxide was not the terminal degradation product. The rate of loss of parent peptide did not vary with pH in the range 6–8.1. The rate of loss was observed to accelerate with pH beyond this range.

Li et al. (90) found the second-order rate constants for the degradation of His-Met in the ascorbic acid–Fe(III) pro-oxidant system show a maximum at pH 6.4. The appearance of a maximal pH was attributed to competing effects of pH on ascorbic acid. Deprotonation of ascorbate at a higher pH (pK_1 = 4.1) facilitates electron donation to Fe(III) and accelerates the initiation reaction, while at the same time ascorbate becomes a better oxygen radical scavenger, inhibiting the reaction. The buffer species also seems to play a role in the kinetics of degradation. In buffers of equal ionic strength, methionine oxidation was faster in the presence of phosphate than in the presence of Tris or HEPES. Phosphate buffers may facilitate the electron transfer from Fe(II) to oxygen, promoting the reaction (90). Buffer species such as Tris or HEPES have a weak affinity for metal ions (91) and result in methionine oxidation reaction rates that are somewhat less than that of phosphate. Tris and HEPES have also been reported to be scavengers of hydroxyl radicals (92), which would be expected to further inhibit reactions in which the hydroxyl radical is the primary reactive oxygen species. In temperature studies, the energy of activation was found to be 23.9 ± 2 kJ/mol, but it is unknown whether this characterizes the formation of the oxidizing species, the oxidation of methionine, or both.

d. Primary Sequence

Li et al. (90) also studied of the effect of primary structure on methionine oxidation. When in a terminal position, Met-Gly-Gly or Gly-Met-Met, the first order degradation rate constants are greater than that of the midposition Gly-Met-Gly. The inclusion of histidine in His-Met greatly accelerates the degradation of methionine. The greatest degradation rates are observed in His-Gly-Met and His-Pro-Met, where methionine is separated by one residue from histidine. Even His-Gly-Gly-Gly-Gly-Met shows enhanced degradation rates (by a factor of 5) compared to Gly-Gly-Met. Whether this is related to the metal ion binding and localized oxidation is not yet known with certainty. The authors do note that the degradation products of these reactions have not been characterized, and complete reaction mechanisms are not yet available.

2. Histidine

Histidine is also highly susceptible to oxidation, either by photocatalyzed or metal-ion-catalyzed mechanisms. Photooxidation of proteins in vivo has been extensively studied (93,94). Photosensitizing agents such as methylene blue (95) or rose bengal (96) are required for photooxidation to take place via the production of singlet oxygen ($^1O_2$). A cycloperoxide ring is produced by the addition of $^1O_2$ to the imidazole ring of histidine (97). Kinetics of photoreactions are often very complex (98), being further complicated by issues of histidine accessibility to solvent and $^1O_2$ (99) as well as simultaneous metal-catalyzed oxidation (100). A variety of products are produced, including the amino acids aspartic acid and asparagine (57,101).

From the standpoint of pharmaceutical formulations, recent evidence suggests that the oxidation products of ascorbic acid (a frequently employed antioxidant) may be potent photosensitizing agents, enhancing histidine oxidation in proteins. Ortwerth et al. (100) reported $^1O_2$ concentrations in the millimolar range and H_2O_2 in the micromolar range after one hour of irradiation with ultraviolet light in the presence of dehydroascorbate and diketogulonic acid (by-products of ascorbic acid oxidation). Complete protection against photooxidation can be attained by protection from light or removal of all dissolved oxygen gas (93).

There has been some study of the effect of primary structure on the photooxidation of histidine. Miskosky and Garcia (93) found little difference in the rate of photooxidation of histidine as the free amino acid and in dipeptides His-Gly and Gly-His. Changing the solvent to acetonitrile/water (1:1) resulted in an order-of-magnitude decrease in the rate of oxidation in all three substrates, suggestive of polarity effects on the rate of reaction.

Histidine appears to be particularly sensitive to transition-metal-catalyzed oxidation, presumably because it often forms a metal binding site in proteins (102). Fenton chemistry at the bound metal ion (such as that as in Eq. 5) could result in high localized concentrations of reactive oxygen species. Histidine residues at the N-terminus appear to be especially susceptable to site-directed metal-ion-catalyzed oxidation (103). Metal-catalyzed oxidation of histidine results in the production of 2-oxo-imidazoline (Fig. 4) (104).

By-products of the proposed reaction include aspartic acid. In the metal-ion-catalyzed oxidation of polyhistidine, the production of aspartic acid is accompanied by scission of the histidyl peptide bond (101), but it remains unclear whether the scission is a part of the reaction mechanism or merely reflects the instability of the 2-oxo-imidazoline ring to the conditions of isolation and analysis (101,105). Chain scission is not frequently observed in proteins upon histidine oxidation.

3. Cysteine

Metal-ion-catalyzed oxidation of cysteine residues usually results in the formation of both intra- and intermolecular disulfide bonds (106,107). Further oxidation of the disulfide results in sulfenic acid.

The mechanism may be summarized as follows (107):

Formation of thiyl radical

$$RS^- + M^{+n} \Rightarrow RS\cdot + M^{+(n-1)} \tag{6}$$

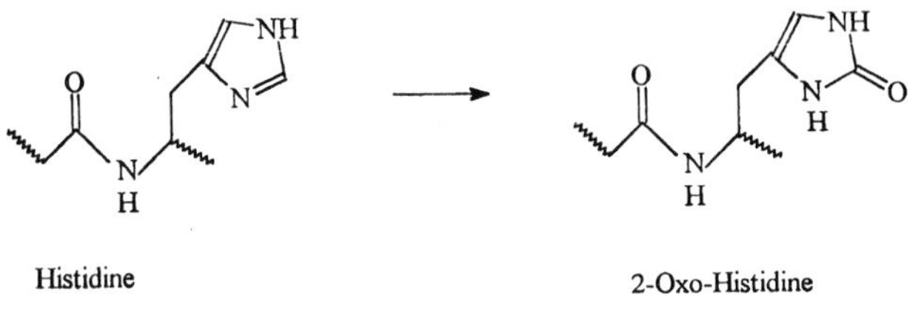

Fig. 4 The first product of histidine oxidation, 2-oxo-histidine.

Formation of disulfide radical anion

$$RS^- + RS\cdot \Rightarrow RS^- \cdot SR \tag{7}$$

Formation of superoxide

$$RS^- \cdot SR + O_2 \Rightarrow RSSR + O_2^{\cdot -} \tag{8}$$

Generation of peroxide

$$RSH + O_2^{\cdot -} + H^+ \Rightarrow RS\cdot + H_2O_2 \tag{9}$$

Regeneration of M^{+n}

$$M^{+(n-1)} + O_2 \Rightarrow M^{+n} + O_2^{\cdot -} \tag{10}$$

This mechanism can result in the production of reactive oxygen species capable of further oxidative damage to the disulfide as well as to other residues in the vacinity. When metal ions are made unavailable by chelation with EDTA, cysteine oxidation is eliminated (108). In general, a pH of 6 appears to be optimal for the oxidation of cysteine in proteins (109). At low pH, the protonation of the sulfhydryl (pK_a 8.5) inhibits reaction with the metal ion (Eq. 6). In the alkaline region, electrostatic repulsion of two ionized cysteines is thought to result in an increased separation of the two residues and a reduced reaction rate (Eq. 7). Oxidation in the absence of a nearby thiol has also been observed (110).

4. Tryptophan

Tryptophan is well known to be a target of reactive oxygen species superoxide (111), singlet oxygen (112), hydroxyl radical, (113), and peroxide (114). The most prominent reaction products of tryptophan oxidation appear to be N'-formylkynurenine and 3-hydroxykynurenine (115). Monohydroxyl derivatives of tryptophan at the 2, 4, 5, 6 and 7 positions have also been observed (Fig. 5). N'-formylkynurenine may be also formed by photooxidation (116). Metal

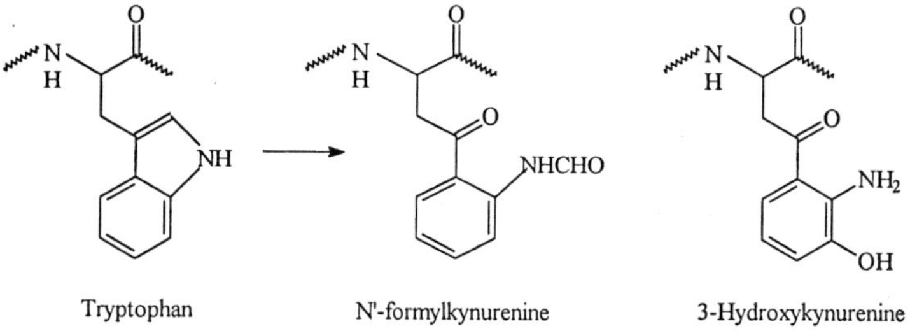

| Tryptophan | N'-formylkynurenine | 3-Hydroxykynurenine |

Fig. 5 Tryptophan oxidation products.

ion catalysis of oxidation appears to play a role in the photolytic mechanism (115).

Very little work has been directed toward an understanding of the influence of primary sequence upon tryptophan photooxidation (117,118). It is known that inclusion of Trp in a peptide bond significantly reduces photocatalyzed radical yield (119). At neutral pH and in the presence of dissolved oxygen, it has been observed that Gly-Trp photooxidizes at a rate approximately 10-fold that of Trp-Gly. Similarly, Leu-Trp degrades at a rate approximately three-fold greater than Trp-Leu. In tripeptides, Gly-Trp-Gly degrades more rapidly than Leu-Trp-Leu. The mechanistic basis of these observations is not clear (117), and additional work remains to be done (118). Under anaerobic conditions, photolytic degradation rate of tryptophan in peptides is also observed, but it is slowed considerably from the rates observed in the presence of oxygen. In addition, Leu-Trp-Leu exhibits even greater stability over Gly-Trp-Gly. These data indicate that both in the presence and absence of oxygen, leucine (with its large side chain) occupying the C-terminal position next to Trp tends to decrease the degradation rate. Although these data do not provide a sufficient base for a generalized rule, it can be speculated that steric effects have an influence on the rate of photodegradation (118).

Photooxidation of Trp in proteins is known to be directly dependent on the accessibility of the residue to oxygen and solvent water (120–122). Trp residues buried in the core of the protein are less rapidly oxidized than those located at the surface of the molecule (122). Micellar solubilization of hydrophobic peptides appears to protect only the Trp residues located in the core of the micelle (121).

5. Phenylalanine and Tyrosine

In the presence of copper ion, phenylalanine is oxidized to 2-, 3-, or 4- (tyrosine) hydroxyphenylalanine (123), as shown in Fig. 6. Tyrosine may photo- or radio-oxidize to 3,4-dihydrophenylalanine (124), or cross-link with another tyrosine to form dityrosine (125). The latter product may be protease resistant and stable to acid hydrolysis (125). Intermolecular cross-linking would result in increased molecular weight of the reaction product.

6. Proline

Hydroxyl radical oxidation by the hydroxyl radical of proline (126,127), as well as glutamic acid and aspartic acid (128), is characterized by site-specific cleavage of the polypeptide chain on the C-terminal end of the residue.

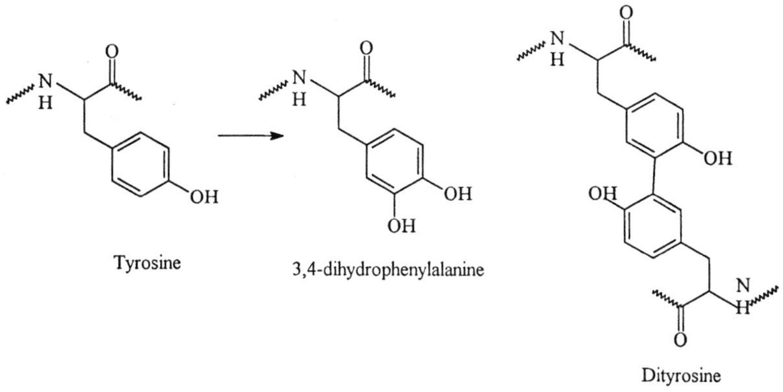

Fig. 6   Tyrosine oxidation products.

D.   Formulation Factors and Oxidation

1.   Polyethers

Many formulations take advantage of the ability of polyethylene glycol (PEG) and PEG-linked surfactants to stabilize proteins against aggregation and thermal denaturation. The potential ability of these adjuvants to promote oxidation becomes an important consideration. PEG and nonionic polyether surfactants are known to produce peroxides upon aging (129). These peroxides are responsible for drug degradation (130,131) in polyether-containing systems. The oxidation of one such polyether surfactant, polysorbate 80, has been shown to release formaldehyde, a potent protein cross-linking agent (132). Careful purification of PEG-containing adjuvants prior to formulation should minimize this potential degradation mechanism.

2.   Sugars and Polyols

Sugars are often employed as lyoprotectants and as part of the vehicle in the formulation and administration of drugs. Literature reports indicate that moderately high concentrations of sugars and polyols seem to inhibit the oxidation of proteins, possibly by serving as hydroxyl radical scavengers (133–135). It was reported in 1996 that various pharmaceutically acceptable sugars and polyols (glycerin, mannitol, glucose, and dextran) were successful in inhibiting ascorbate–Cu(II)-induced oxidation of the protein relaxin and of model peptides Gly-Met-Gly and Gly-His-Gly (136). Results of experiments with glycerin show that contrary to expectations, the protective effect is not the result of radical scavenging. Rather, these authors concluded that the protective effect of the sugars and polyols was due to complexation of transition metal ions. This is in accord with reports indicating weak, but stable complexes of metal ions with sugars (137,138). Production of reactive oxygen species can be diminished or eliminated by means of competition with the peptide for binding of the metal ion (Eq. 5). Dextran also inhibited oxidation, but a detailed mechanism of the protective effect was not given. The safety and availability of these inexpensive additives make them very attractive as agents protective against oxidation.

The inclusion of glucose in a protein formulation is not without potential risks. Glucose has been shown to participate in oxidation reactions catalyzed by metal ions (139,140). Methionine oxidation products have been observed (141). Upon reaction of glucose with Fe(II), an enediol radical anion intermediate is formed which quickly reacts with molecular oxygen to form the ketoaldehyde (which itself can react with a free amino groups on the protein, forming a keto aminomethylol) and the superoxide radical. As a reducing sugar, glucose has been shown to covalently modify relaxin by adding to the side chains of lysine and arginine and by catalyzing the hydrolysis of C-terminal serine amide bond. Neither nonreducing sugar (trehalsoe) nor polyhydric alcohol (mannitiol) participates in these reactions (136).

3. Antioxidants

Antioxidants are commonly employed to protect both small-molecule and peptide/protein drugs from oxidation in pharmaceutical formulations. These are essentially sacrificial targets that have a great tendency to oxidize, consuming pro-oxidant species. The choice of an antioxidant is complicated in proteins and peptides because of the interaction chemistries possible between the antioxidant and the different amino acid side chains (142). Even antioxidants that are themselves benign to proteins can become potent pro-oxidants in the presence of trace amounts of transition metal ions (e.g., ascorbic acid). In the absence of metal ions, cysteine, as a free amino acid, may act as an effective antioxidant (4,107,134). By virtue of its singlet oxygen scavenger activity, α-tocopherol has shown protective effects against photooxidation of proteins within lipid membranes (143). Whether this additive is effective in reducing oxidation of proteins or peptides in lipid-based delivery systems, such as liposomes or emulsions, remains unknown.

4. Processing and Packaging

Removal of oxygen from solution by degassing processes may be an effective means of inhibiting oxidation in protein and peptide solutions (142,144). Even very low concentrations of oxygen in the headspace will promote oxidation (145). To minimize foaming in protein solutions during degassing, Fransson and coworkers have suggested cyclic treatments of low temperature and low pressure, followed by exposure to atmospheric pressure nitrogen gas (64). Packaging in a light-resistant container may be helpful in reducing light-cata-lyzed oxidation. It should be kept in mind that glass may release minute quanti-ties of metal ions sufficient for metal-ion-catalyzed oxidation (146).

5. Lyophilization

The influence of moisture content on oxidation and other protein degradative reactions has been explored by a number of authors (4,147,148). Most often, residual moisture enhances the degradation of proteins (4). Hageman has listed both oxidation promotion and oxidation inhibition mechanisms of water (148). Pro-oxidant activities of moisture is believed to include mobilization of cata-lysts, exposing new reaction sites by swelling, and decreasing viscosity of the sorbed phase. The oxidation promotion activities of water are thought to be initiated at or near monolayer coverage, where conformational flexibility of the protein is enhanced (147). Oxidation inhibition activity of sorbed water is thought to arise from retardation of oxygen diffusion, promotion of radical recombination, decreased catalytic effectiveness of transition metals and dilu-tion of catalyst (148). Much higher water content is required for the modest oxidation inhibitory effects to become manifest (148). The existence of both pro- and antioxidant effects of moisture would be consistent with the widely varying experimental results observed. For example, Fransson et al. (64) have observed no dependence on moisture of the second-order rate constants for methionine oxidation in insulin-like growth factor, while Pikal et al. (144) found a strong dependence on moisture for methionine oxidation in human growth factor. Clear-cut mechanistic interpretation of these differing oxidation results is not yet possible because of the possible masking effects of formula-tion additives and moisture-dependent protein conformational state. Residual moisture values in lyophilized proteins of less than 1% tend to be associated with enhanced stability of the protein upon storage (147). In production lots, removal of water to attain such low levels of residual moisture is quite expen-sive. As a cost-effective alternative, Pikal et al. have suggested the inclusion of an amorphous excipient to act as a sink for residual water to enhance the stability of lyophilized proteins (145).

III. PHYSICAL CONSIDERATIONS

The physical or conformational stability of proteins (i.e., changes in tertiary or higher order structure), and thus protein function, can be affected by a number of environmental parameters. Relatively small changes in temperature, pressure, pH, concentration of denaturing agents (e.g., guanidine hydrochloride, surfactants), or exposure of the macromolecule to mechanical disruption can lead to an irreversible loss of protein function. The primary mechanism for protein inactivation by these agents or processes involves the denaturation or unfolding of the protein macromolecule. Protein unfolding refers to the loss of tertiary structure and the formation of a disordered protein in which the proper intramolecular contacts within the protein no longer exist. The intermolecular recognition events necessary for proper protein folding are usually cooperative and reversible (as discussed below) upon removal of the denaturing agent. However, unfolding can be followed by secondary irreversible inactivating processes such as the chemical changes described at the beginning of this chapter, or by other physical processes such as aggregation of the protein to form higher order oligomers or ''aggregates'' and adsorption to surfaces. Therefore, the preservation of protein tertiary structure becomes paramount to preventing losses in protein function. Although scientists have made significant progress in understanding and predicting protein folding, the complex interplay of the molecular determinants that drive intraprotein recognition events makes protein stabilization an interesting challenge for the protein formulator.

The tertiary structure of proteins is driven by two classes of noncovalent interactions, electrostatic and hydrophobic. Electrostatic interactions include ion pairs (149), H bonds, weakly polar interactions, and van der Waals forces. ''Hydrophobic interactions'' refer to actions and hydration effects of nonpolar groups. The interplay between enthalpy and entropy makes individual changes in a physical parameter complex, and the exact causes are still controversial. In this section we attempt to sort out experimentally accessible parameters that may be useful to the protein formulator for conducting preformulation studies, formulation of the protein for a desired extent of stability, and ways to test for stability.

A. Thermodynamics of Protein Stability

1. Thermodynamics of Protein Folding

The simplest model used to describe reversible protein stability is a two-state model in which an equilibrium exists between the native (N) and denatured states (D).

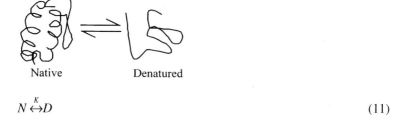

Native   Denatured

$$N \overset{K}{\leftrightarrow} D \qquad\qquad (11)$$

Under physiological conditions, the native and denatured states are in equilibrium. The equilibrium constant for unfolding is thus given as follows:

$$K = \frac{[D]}{[N]} \qquad\qquad (12)$$

The change in Gibbs free energy, regardless of any intermediate complexes, can then be deduced from the equilibrium constant,

$$\Delta G = (G_D - G_N) = -RT \ln K \qquad\qquad (13)$$

where R is the gas constant and T is the absolute temperature. The free energy for the equilibrium is typically 5–20 kcal/mol, reflecting a remarkably low stability of the native protein (150,151). The temperature coefficient of the equilibrium constant is given by,

$$\frac{d \ln K}{dT} = \frac{\Delta H}{RT^2} \qquad\qquad (14)$$

which is commonly referred to as the van't Hoff equation. The change in Gibbs energy as a function of temperature is given as follows:

$$\Delta G = \Delta H - T\Delta S \qquad\qquad (15)$$

where ΔH and ΔS are the changes in enthalpy and entropy, respectively, at a given temperature. Equation 15 reveals that the free energy is a delicate inter-

play between the enthalpy and entropy changes, which are typically large numbers (e.g., 50–200 kcal). Since the small changes in free energies associated
with the transition from the native to the denatured state are the difference of
large numbers, this makes the difference in free energy very sensitive to small
perturbations in the attractive and repulsive intramolecular interactions. This
makes protein stability easy to manipulate but makes exact predictions of protein stability difficult. The dependency of enthalpy and entropy on temperature
can be described in terms of the heat capacity (ΔC_p) at a constant pressure:

$$\Delta H = \Delta H^\circ + \Delta C_p(T - T_0) \tag{16}$$

$$\Delta S = \Delta S^\circ + \Delta C_p \ln\left(\frac{T}{T_0}\right) \tag{17}$$

where ΔH° and ΔS° are the enthalpy and entropy at a given reference temperature, T°. The change in heat capacity, ΔC_p, is the difference in heat capacity
between the N and D states and, in most cases involving proteins, is large and
relatively constant within experimental error (152). The large value for ΔC_p
largely reflects the restructuring of solvent upon protein unfolding (153,154).
Robertson and Murphy have summarized the thermodynamic variables for
globular proteins of known structure (155). Combining Eqs. 14, 15, and 16,
yields Eq. 18, which is the modified Gibbs–Helmholtz equation (for an excellent review, see Ref. 155).

$$\Delta G = \Delta H^\circ - T\Delta S^\circ + \Delta C_p(T - T_0 - T \ln\left(\frac{T}{T_0}\right) \tag{18}$$

It is important to remember that Eq. 18 holds true only for the two-state model,
in which no stable folding intermediates exist. This assumption has been
shown to be valid for many small globular proteins, although exceptions do
exist (156). In some of these cases, the existence of intermediates that are not
significantly populated or are short-lived cannot be ruled out.

The value of ΔG is thus the fundamental measure of the stability of the
protein. In cases that allow the assumption of a two-state model, the ΔG can
be determined from measurements of K as a function of temperature by means
of spectroscopic methods (bioassays, chromatography, electrophoresis, etc.)
to determine relative populations of the native and denatured proteins (157).
At the melting temperature T_m, ΔG is zero and by Eq. 5, ΔS is equal to $\Delta H/T_m$. Therefore, Eq. 18 can be used to fit the experimental values of data on
$\ln K$ versus T, to extract the values for ΔH_m, T_m, and ΔC_p at T_m. Once these

parameters have been determined, ΔH and ΔS can be calculated over a wide range of temperatures. Typically, a maximum in stability is observed between 0 and 25°C (158) (to be discussed further). However, the range of temperatures in which the protein is stable will change depending on the presence and concentration of other agents.

2. Reversible Versus Irreversible Protein Denaturation

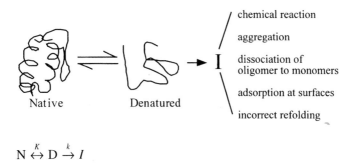

$$N \overset{K}{\leftrightarrow} D \overset{k}{\to} I$$

Reversible denaturation or unfolding can be followed by processes that create irreversibly inactivated forms of the protein (I) (159). While K is the equilibrium constant for reversible denaturation, which correlates to a protein's thermodynamic stability as described above, k is the rate constant for the irreversible inactivation of the protein, which is a measure of a protein's long-term stability (160). While the diagram shows inactivation after denaturation only, inactivation can proceed directly from the native state, although the former is typically observed. Unfolded proteins are more prone to proteolysis than tightly packed, globular proteins (161), often tend to aggregate into insoluble masses, and are able to become kinetically trapped in improperly folded conformations. Inactivation can sometimes be prevented, and some methods are discussed in the subsections that follow.

B. Physical Protein Stability in Solution

Pharmaceutical formulations are complex systems, and it is often difficult to separate the effects of any single variable that can account for a shift in the equilibrium to favor the denatured protein state. Losses in protein physical stability most likely are due to an interplay between many different stabilizing [e.g., hydrogen bonding and the hydrophobic effect (162)] and destabilizing [e.g., configurational entropy (163)], effects. A large range in each physical

parameter exists to which proteins have adapted: for temperature, -5 to
$110°C$; water activity, $1-0.6$ [corresponding to $\leq 6M$ salt (164)]; pH, $4-12$.
Although alteration of these parameters often leads to a reversible denatured
state, it is important to remember that unfolding is usually the first step in
irreversible losses in protein structure, and thus keeping the protein folded is
paramount in stabilizing protein structure.

1. Temperature

The stability of proteins to thermal stress is an important variable to the protein
formulator. Changes in temperature may accompany processing [e.g., lyophilization (165) or spray drying] and serves as a convenient thermodynamic variable to probe the stability and shelf life for the protein pharmaceutical through
accelerated protein stability testing. Furthermore, calorimetry can be used to
obtain accurate measurements of the thermodynamic properties of unfolding
as a function of temperature.

   Thermally induced unfolding is highly cooperative and often reversible
(166,167). Protein stability curves (168) refer to plots of ΔG versus temperature and can be described by the Gibbs–Helmholtz equation (Eq. 18), which
can be written as follows:

$$\Delta G(T) = \Delta H_m\left(1 - \frac{T}{T_m}\right) - \Delta C_p\left[T_m - T + T\ln\left(\frac{T}{T_m}\right)\right] \qquad (19)$$

where $\Delta G(T)$ is the ΔG at temperature T, T_m is the midpoint of the thermal
unfolding curve, and ΔH_m is the enthalpy change for unfolding measured at
T_m. At T_m, $\Delta G = 0$ and ΔS has just been replaced by $\Delta H/T_m$. The temperature
for maximum stability (T_s) occurs at the temperature when $\Delta S = 0$, and is
given by

$$T_s = T_m\exp\left(-\frac{\Delta H_m}{T_m\Delta C_p}\right) \qquad (20)$$

Typically T_s is between -10 and $35C°$ for most proteins. Figure 7 is a plot
of ΔG versus temperature for the protein RNase T_1 at pH 7 (169,170). Equation
19 was used to calculate the solid line, with the parameters listed in the figure
legend. The plot illustrates that minimal changes in temperature, by heating
or cooling, can have profound effects on protein folding. Although the folding
is usually reversible, partial unfolding frequently represents the first step in
thermal deactivation (171,172). Also, once partial unfolding has occurred, the

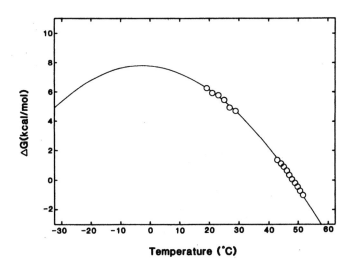

Fig. 7 ΔG as a function of temperature for the unfolding of Rnase T1 and pH 7.0 (30 mM MOPS). The points in the 19–30°C temperature range are from urea denaturation experiments, while those above 40°C are from thermal unfolding curves and Eqs. 11, 12, and 13. The solid curve was calculated from Eq. 19 using $T_m = 48.25°C$, $\Delta H_m = 95.2$ kcal/mol, and $\Delta C_p = 1.72$ kcal/mol deg. (Reprinted with permission from Biochemistry 1989, 28, pg. 2524. Copyright 1989 American Chemical Society.)

protein is highly susceptible to irreversible denaturation processes such as aggregation and incorrect folding schemes, which severely impact long-term protein storage and function.

Much work in the area of protein stability with respect to temperature has been derived from an interest in understanding how psychrophilic (173) · and thermophilic organisms survive (for reviews Spanning the 1990s, see Refs, 174–177). There is a strong effort in the development of thermally stable proteins and understanding how subtle changes in the amino acid structure can significantly alter protein stability. Primary sequence analysis of proteins belonging to thermophyllic organisms reveal several consistencies.

First, the positively charged residue of choice appears to be arginine over lysine. Reasons for such a substitution in heat-stable proteins include arginine's higher pK_a, which would prevent this amino acid from losing its charge at higher temperatures (pK_a decreases with increasing temperature) (178), the increased surface area of arginine, and its shorter hydrocarbon attachment to the polypeptide backbone (160). There is a decrease in asparagine content, which is especially susceptible to deamidation at high temperatures and neutral pH. Finally, an increase in proline content appears to correlate with increased thermal stability, likely because of the increase in rigidity in protein structure (179). However, enhancement of protein rigidity to increased stability may come at a cost, since protein flexibility is required for protein function. On the other hand, surveys of the psychrophilic organisms reveal only minor changes in the primary structure. However, significant changes in intraprotein interactions are significantly reduced, presumably to preserve protein flexibility (173). The ease with which protein engineers can manipulate primary protein structure makes enhancing protein thermal stability by small changes in amino acid sequence a feasible mechanism.

Many of the proteins belonging to thermophilic organisms have tertiary structure (180,181) similar to that of their nonthermophilic counterparts. However, it has been observed that thermophilic proteins have additional salt bridges, hydrogen bonds (182), and hydrophobic interactions (180). Of these interactions, the major driving force that serves to enhance protein thermal stability appears to come about by an increase in the hydrophobic interactions of the hydrophilic core (172,183). Increases in the overall hydrophobic character of the thermophilic protein are brought about by loss of surface loops, increase in helix-forming amino acids, and restriction of N-terminal residues (164,177). This enhancement of stability is believed to come from a large increase in the enthalpy of unfolding, which increases with increasing hydrophobicity (184,185). The enthalpic increase is attributed to the melting of water cages that surround the exposed nonpolar side chains (171).

In addition to optimization of the protein's structure, nature protects proteins against thermal stress, in particular via cryoprotection, by introducing free amino acids (186), organic salts (e.g., trehalose, sucrose), and polymers such as polyethylene glycol (164,187–189). The low molecular weight compounds are hypothesized to be excluded from the protein and thus to increase water activity around the protein. However, these additives may also operate through a mechanism whereby the chemical and physical processes are kinetically hindered by the high viscosity of the additives. The reader is referred to Sect. IV, which specifically addresses stabilization of proteins with osmolytes. Addition of stabilizers should be considered carefully, since both reduced and elevated temperatures can cause the crystallization and inactivation of added excipients that might hinder their stabilizing properties (190,191). Therefore, independent from the thermal stability of the protein, the formulator needs to consider the stability of the entire formulation to changes in temperature.

Last, the long-term storage temperatures for proteins has been related to the glass transition temperature (T_g). At this temperature an equilibrium exists between protein in a glassy and a rubbery state. Additives that raise T_g have been incorporated with some success into formulations to decrease molecular mobility in lyophilized protein products (192–195). A common misconception is that molecular mobility below T_g is sufficiently low to forbid large-scale structural changes. Many direct experiments using solid state NMR spectroscopy have shown that there is significant mobility of molecules at or near T_g (196,197). It has been shown experimentally as well that storage of proteins at or just below T_g is not sufficient to prevent physical and chemical changes of the protein (198).

2. pH

Proteins are typically most stable (and often least soluble) at their isoelectric points (pI), where opposing charges serve to stabilize the protein structure (149,199,200). Figure 8 is a plot of the free energy [denoted as $\Delta G(H_2O)$ as a function of pH] for RNase T1 (169). The maximum in stability occurs at a pH where the net charge on the protein is zero. Acid or base environments can influence protein stability primarily through changes in the electrostatic

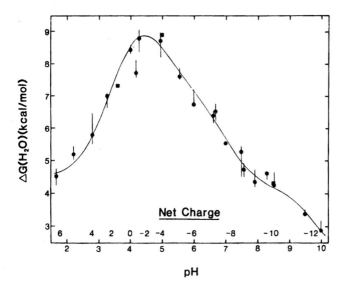

Fig. 8 The free energy for unfolding $\Delta G(H_2O)$ for Rnase T1 plotted as a function of pH. An estimate of the net charge is given at each pH based on titration studies (326). The solid line has no theoretical significance. (Reprinted with permission from Biochemistry 1990, 29, pg. 2567. Copyright 1990 American Chemical Society.)

free energy of the protein as the ionization states of acidic and basic groups of the protein are changed. Through the use of site-directed mutagenesis, it was found that acid and base denaturation is primarily caused by buried charges rather than by changes in the overall surface charge–charge repulsion (201–204) and that the electrostatic free energy difference between native and denatured states is brought about through changes in the pK_a values of a relatively small number of amino acids (170,205–207). Acid denaturation of a protein below its pI is governed by a net increase in positive charges that, as pH is lowered, can generate repulsive forces that can exceed the stabilizing forces and lead to the loss of the native state (208). Furthermore, changes in the pK_a values of the ionizable groups in the native and denatured state (199) can further disrupt the native structure. NMR studies of RNase T_1 have shown that certain ionizable groups have higher pK_a values in the folded state, so that decreasing pH would allow the functional groups in the native state to bind protons more tightly (171), shifting the equilibrium to the native state. Similarly, losses in protein stability are observed as the protein environment pH is further lowered as the pK_a aspartic acid residues is shifted to higher values in the unfolded conformation so that the unfolded becomes the most thermodynamically stable state (209). Various empirical methods may be employed to determine the pK_a of the ionizable groups in proteins. Ultraviolet, infrared, NMR, and fluorescence spectroscopies and capillary electrophoresis (210) have all been used to obtain these values.

The theoretical treatment of pH dependence of protein stability had led to powerful modeling studies that allow the accurate prediction of the stability of a protein to pH changes. These calculations rely on the electrostatic calculations to determine the pK_a of each ionizable group on the protein (200,211–213), but they require an accurate representation of the protein structure in native and denatured states. This allows for the calculation of the titration curve for the unfolding free energies as a function of pH. These methods have been found to be accurate enough to provide a useful tool in the interpretation of experimental results (200,211).

3. Solvation

Ions can interact with proteins in a variety of ways and can affect protein stability dramatically. Many empirical studies exists for solvation-induced perturbation of protein structure [for a review, see Arakawa et al. (214)]. In general, protein structural stability in the solution is associated with agents that are preferentially excluded from the native protein; that is, in the presence of these molecules, proteins in the folded conformation are preferentially hydrated. Denaturants, in contrast, are preferentially bound to the denatured protein. The stabilizing, destabilizing, precipitating, and solubilizing properties of these agents are manifested through affecting the balance between the affinities of the proteins for water and the agent (215).

a. Salting-In/Salting-Out

Increasing the solubility of a protein with low salt concentrations, salting-in, is due to Debye–Hückel screening. At these concentrations, the ion interactions with the protein are nonspecific and electrostatic. At higher ionic strengths (typically >1.0), protein salting-out dominates, which is linearly related with increasing ionic strength (215) and is relatively ion specific (216). In general, salting-out is hypothesized to be due to a preferential exclusion of the solute from the protein domain, which results in the changing in the hydrogen-binding properties of the water (217). However, the exact molecular mechanisms are not understood (218). As a result, predictions of how molecular interactions affect protein stability tend to be protein specific and dependent on the system conditions such as temperature and counterion identity, (i.e., size and charge).

b. Stabilization/Denaturation

Organisms in nature face environmental stress, (viz., freezing or osmotic stress) by accumulating "compatible" organic solutes that stabilize protein structure. These cryoprotectants or osmolytes are low molecular weight molecules that are produced by the living organisms in response to stress (219). Studies of sugars (sucrose, glucose, lactose, trehalose), polyhydric alcohols (mannitol, ethylene glycol, sorbitol, xylitol, inositol), and glycerol, which are associated with the stressed organisms, showed that such solutes stabilize the native structure of proteins by means of their preferential exclusion from the surface of proteins (159,214,220–223). This preferential exclusion could be due to steric, solvophobic effects and/or increases in surface tension (224–226). Baskakov and Bolen have shown the extraordinary stabilization ability of some osmolytes which have been illustrated to be able to force thermodynamically unstable proteins to fold (227).

The exclusion of the solutes from immediately outside the domain of the protein causes a preferential enrichment of water around the protein (228) (see Fig. 9). The result is that these additives will increase the chemical potential of the protein. This increase in the free energy of the system is proportional to surface area and is thus more unfavorable for the unfolded state (Fig. 9), which has a greater surface area exposed to solvent. This shifts the equilibrium back to the folded, native state (214). Changing protein hydration due to addi-

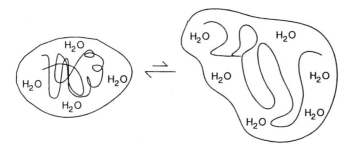

Fig. 9 Schematic representation of preferential hydration of proteins. The greater exposed surface area in the denatured state shifts the equilibrium to the left, favoring the native state. (Adapted from Ref. 215.)

tives is one of the simplest approaches to stabilizing proteins (229). Since, however, not all additives that cause preferential hydration work to stabilize proteins under all conditions (different temperatures, pH, etc.), a thorough study of the dependence of the additive being used for stabilization on solution conditions must be made (230). A classification has been established that attempts to accurately describe which additives can be classified as stabilizers (220,221,230). The first class includes solutes for which the preferential hydration of the protein is independent of the solution conditions (i.e., pH, solute concentration), while the second class includes solutes that vary with conditions. Solutes belonging to the first class stabilize proteins, while the others do not always do so. Therefore, thorough measurements (231) of the preferential hydration of the native state of a protein should be performed to verify a molecule's stabilizing action.

Conformational stability can also be increased by preferential binding of ions to a folded molecule (159,172). In the case of RNase T1 (232), there exists a cation and anion binding site on the surface of the protein. As might be expected, this can be used to the advantage of protein engineers, who can engineer into the native protein binding sites to favor the folded protein. Furthermore, advances in performing theoretical calculations are allowing accurate predictions of changes in protein stability after mutation (233).

Denaturation of proteins by chaotropic agents such as guanidine hydrochloride and urea are brought about by direct specific interaction of the small molecule with the protein. Experimental evidence from thermodynamic equilibrium and calorimetric titration techniques has shown that unfolding is favored because a favorable free energy gain is produced upon interaction of the denaturants with residues in the protein that become exposed on denaturation (158,168,215,234,235). In some cases, the denaturing solute can be repelled by charges (236) of the protein in the folded state but, as the protein unfolds, the charges are dispersed and the solute interacts with the newly exposed non-polar residues (237). As mentioned above, salting-out sometimes can be followed by denaturation. In these cases, increases in ionic strength are most likely inducing denaturation by decreasing attractive charge–charge interactions (200).

4. Surfactants

The interaction of sodium dodecyl sulfate (SDS) with proteins is well known to cause the unfolding of proteins and is used commonly in gel electrophoresis to disrupt tertiary and secondary structure. This disruption of structural order occurs after the binding of SDS to the protein (238,239). Experiments with soybean trypsin inhibitor (240) have shown that only a few SDS molecules are needed to perturb the protein's structure and that the loss of structure gradually increases as additional SDS is added. After binding 7–10 molecules of SDS, the protein displays only a limited amount of α-helical order. SDS's long aliphatic tail is believed to penetrate hydrophobic pores, while shorter alkyl sulfates cannot. Charge–charge interactions between the negatively charged SDS and positively charged sites are also presumed to play a role in this interaction, in as much as SDS has been shown to have no destabilizing effects on proteins with large negative charge densities (241).

In contrast to the SDS–protein interactions, surfactants are commonly being exploited for their stabilizing properties (e.g., to increase refolding yields in protein purification) (242). Protein formulations also commonly contain polymers, polyols, and nonimic and anionic surfactants as stabilizing agents. The bulk of protein–surfactant interactions are not well understood. Some mechanisms that may prevent losses in protein stability include prevention of surface-induced deactivation of proteins (243,244) and inhibition of aggregation and precipitation (242,245–247). In all of these cases, it is believed that the surfactant binds the protein and reduces the protein's available hydrophobic surface area, thus reducing the protein's self-association and any deleterious interactions with nonspecific hydrophobic surfaces.

Techniques used to determine surfactant–protein stochiometry include surface tension (248), viscosity (249), dye solubilization (239), dialysis (250), ion-selective electrodes (251–253), electron paramagnetic resonance (EPR) spectroscopy (254), and analytical ultracentrifugation (255). These measurements have shown that protein–surfactant aggregates form well below the critical micelle concentration for the surfactant and that the aggregates are typically smaller than micelles, although some evidence suggests that micelles do form in some systems (256,257).

5. Processing

a. Shear

Protein formulation and manufacturing often employ processes in which the molecule is subject to shear forces. These processes include mixing, flow both in solution and in powder form, filtration, and passage through pumps (258,259). Empirical modeling of data obtained on shearing proteins shows that for many proteins there is a logarithmic relationship between loss in activity and shearing time (260–262). Conformational changes and accelerated aggregation (258) have also been attributed to high shear rates. Burgess et al. have developed a method to determine protein stability to shear by measuring the interfacial shear rheology of adsorbed protein layers and the effects of additives (263).

b. Dehydration and Lyophilization

Destabilization of protein structure by lyophilization may be brought about by several mechanisms. There may be effects due to cold denaturation, concentration/crystallization of salts, changes in pH, and the creation of solid–liquid interfaces if ice forms in the solution. All these topics have been or will be addressed in this section. However, an additional mechanism, dehydration, has been reported for loss of protein structure during a lyophilization (187,264–266). This has been measured using infrared, Raman, and NMR spectroscopies (267–268). Whether the changes are solely due to dehydration, remains somewhat controversial, although overwhelming data suggest that dehydration alone is responsible. All proteins appear to have water associated with the native form. Loss of structure due to drying has been found to be in part irreversible, making avoidance of this process or discovering ways of stabilizing proteins to dehydration matters of serious concern in the manufacture of protein pharmaceuticals (265). When considering a formulation that is going to be lyophilized, the formulator should take precautions for preserving protein structure due to dehydration, (i.e., sucrose or mannitol) as well as cryoprotection (187). The problem of dehydration may be thought to be solved by lyophilizing with residual moisture. However, reports of moisture-induced aggregation document several case studies in which residual moisture had sufficient mobility for noncovalent aggregation as well as chemical reactions (159,269–272).

C. Surface-Associated Mechanisms

1. Liquid–Air Interface

Many proteins have been found to denature at the air–water interface (273–276). This property can have an impact during shipment of solutions of protein pharmaceuticals, during reconstitution of lyophilized protein products, and for their administration or any processing that involves mechanical disruption that could result in foaming of the protein solution (243,275). Once at the interface, protein chains can become entangled as the interface is further perturbed. It has been suggested that primary aggregation can occur at an interface, which can seed the bulk formation of larger insoluble aggregates (277,278). Thus, surface-induced denaturation will often lead to precipitation (273,275).

The driving force responsible for surface accumulation of proteins is the reduction in the interfacial surface energy, which reduces the overall free energy of the entire system. It has been observed that the less stable proteins tend to be more surface active, presumably owing to exposure of nonpolar groups upon denaturation. Changes in the surface tension of a protein solution with time are a direct measure of the affinity of the molecule for the interface and have been shown to be a useful indicator for predicting the stability of proteins in aqueous solutions and in designing formulations that may prevent surface denaturation (275). A systematic study by Wang and McGuire (276) on T4 phage lysozyme showed that among mutants with decreased free energies of unfolding, less stable variants decreased surface tension to a greater extent and occupied more surface area. Levine et al. (275) were also able to correlate the surface tension decrease with the ease of denaturation and precipitation. Surfactants can often aid in stabilization against such denaturation, presumably via preferential adsorption at the denaturing interfaces. Surfactants that have been used include polysorbate 80 and SDS (279).

2. Solid–Liquid Interface

Nonspecific protein adsorption or fouling of surfaces can lead to denaturation of the protein and inactivation of the protein due to irreversible binding (280) and/or subsequent aggregation events (246,259,281). Surfaces for potential inactivation of proteins that are important for consideration in protein formulation include delivery pumps (e.g., infusion sets) (259,282–284), silicone rubber tubing (246,285), glass and plastic containers (e.g., intravenous bags)

(243,245,246,258,286–288) and polymeric surfaces in protein drug delivery devices (285). Undesirable adsorption of protein can also lead to the failure of devices.

Solid surfaces in contact with protein solutions can lead to the denaturation and inactivation of the protein depending on the electrostatic and hydrophobic character of both the surface and the protein (280,285). A variety of mechanisms have been implicated for protein adsorption. Matsuno et al. (289), in a study done with γ-crystalline water-soluble proteins, found that hydrophobic surfaces tend to interact with the internal apolar regions of this protein, thus favoring denaturation. Furthermore, the degree of unfolding increased with surface hydrophobicity: see also Sefton and Antonacci (281). However, this is not the only identified mechanism for protein adsorption. Andrade and Hlady (290) show that the carbohydrate moiety of plasma proteins is responsible for the surface-induced structural instabilities that dominate the adsorption process.

The best way to prevent surface-induced denaturation requires a knowledge of the protein adsorption process and a characterization of the solid surface, including hydrophobicity, flexibility, and porosity (288,290). Polyethylene oxide (PEO) has been shown to resist nonspecific protein adsorption. Possible mechanisms for PEO's inertness are its low interfacial free energy, its water solubility due to a unique hydrogen-bonding network, and its steric stabilizing effects (291). Coatings containing PEO have been employed as a treatment in rendering hydrophobic surfaces protein resistant. Surfactants containing PEO functional groups have been used to treat a variety of commonly used commercial polyethylenes. The treated surfaces were found to be resistant to protein adsorption as studied by X-ray photoelectron spectroscopy (XPS) and through the use of radiolabeled proteins (291). Copolymers of alkyl methacrylates with methoxy (polyethylene oxide) also have been studied for use as coatings or as cleaners for the removal of proteins already adsorbed (292). Protein adsorption onto plastic bags can be reduced by depositing a thin layer of triethylene glycol monoallyl ether. Beyer et al. (293) deposited this layer by means of a plasma-aided manufacturing technique that involves plasma polymerization of the monomer during the deposition process to form on the solid substrate PEO structures that are stable in aqueous solution. Radiolabeled bovine serum albumin adsorption studies with these modified surfaces showed that there was significantly less protein adsorbed to the modified substrate. Preadsorption of a protein layer may also prevent adsorption of the protein therapeutic of interest (280,284). Finally, the use of long chain surfactants has also been shown to interfere with insulin inactivation due to surface-induced instabilities (246).

Another solid–liquid interface frequently encountered by the protein formulator is the ice–water interface that is generated during the freezing of protein solutions. Ice can be formed during routine storage conditions for the protein or through lyophilization protocols. This type of surface-induced denaturation has been reported to correlate with the amount of ice surface area present during the freezing process (294). In this report, the authors report that denaturation induced by the ice–liquid interface can been reduced by the use of surfactants.

3. Immobilization onto a Solid Surface

Immobilization can be used as a protein stabilizing strategy. Much of the research in this area comes from the study of enzyme immobilization (295,296). A number of options can be used to immobilize proteins including covalent chemical linkage (for review see Ref. 297), lipid-mediated immobilization via ligand binding, metal chelation, or electrostatic interaction (for review see Ref. 298), and physical immobilization. In addition, proteins that do not readily adsorb noncovalently to surfaces can be fused with proteins that do (295) without interfering with either protein's activity.

The exact mechanisms for enhancement of protein stability upon immobilization are not well understood. Stabilization may be due to altered protein conformation. A protein that is rigidly fixed on a support may be more difficult to unfold (299). Ribonuclease A covalently immobilized on silica beads was found to have a greater stability to thermal denaturation, which was hypothesized to occur through a decoupling of structural domains which, before immobilized underwent cooperative unfolding (296). Alterations in protein microenvironment may also aid in stabilizing the protein (299).

IV. STABILITY TESTING

A. Thermodynamic Protein Stability

The change in free energy ΔG between the native and denatured states is the fundamental measure of protein stability. Using calorimetric techniques such as differential scanning calorimetry (DSC: excellent reviews can be found in Refs. 300–302), protein folding stability can be measured directly and is therefore model independent. Typically in these experiments, the protein is heated until it undergoes denaturation (although cold denaturation can also be used). The observable parameters are the enthalpy of the unfolding transition, ΔH_m, which is obtained from the area under the curve of the measured excess heat capacity as a function of temperature, and the change in heat capacity between the states, ΔC_p, which is obtained from a shift in the baseline from the native to the denatured states (see Fig. 10). Heat capacity is slightly temperature dependent but is usually assumed to be constant, a practice that does not introduce significant amounts of error (162). An advantage to using calorimetric methods is that the measured ΔH_m can be compared with that determined from a fit to the van't Hoff equation (Eq. 14), which assumes a two-state model; discrepancies indicate the presence of intermediates. Finally, the Gibbs–Helmholtz equation (Eq. 18), can be used to calculate G. The melt temperature T_m, the temperature at which the native and denatured states are in equilibrium, is also used as a measure of protein stability (see Fig. 10). This can be especially useful in comparisons to clarify the factors that may affect protein stability in the presence of additives (191).

Potential problems with this analysis arise in determining the ΔC_p. It is difficult to get accurate measurements of ΔC_p from shifts in the baseline. Typically, careful measurements of the unfolding need to be done and should be performed at several values of T_m, which requires a perturbation of T_m. This is commonly done using pH to shift T_m and then measuring the corresponding ΔH (303). A plot of ΔH as a function of T_m will yield a slope that is the ΔC_p. This assumes that there is no change in ΔH as a function of pH: that is, all

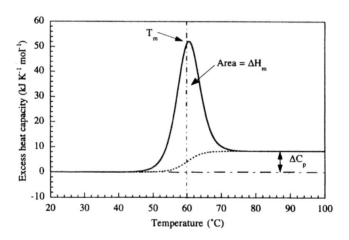

Fig. 10 Simulated differential scanning calorimetry experiment for the two-state unfolding of a globular protein. The simulation assumed the following values: $T_m = 60°C$, $\Delta H_m = 418$ kJ mol^{-1}, and $\Delta C_p = 8.4$ kJK^{-1} mol^{-1}. (Reprinted with permission from Chemical Reviews 1997, 97, pg. 1253. Copyright 1997 American Chemical Society.)

the effects are due to changes in ΔS_m. Care also needs to be taken to be sure that the temperature-induced denaturation is reversible, which is necessary to apply this thermodynamic analysis. Reversibility can be checked by multiple scans of the same sample and by checks to be sure that the change in the measured excess heat is independent of scan rate (155). Other limitations of this technique include the high concentrations (i.e., > 1 mg/mL) and large sample volumes (> 1–2 mL) required for each experiment. With some proteins, these concentrations enter into the range where aggregation is a problem. Additional sources of uncertainty come from the determination of the protein concentration. Typically the reproducibility is on the order of 5%, given a 2% variability in the determination of the extinction coefficient for the protein (304). Experimentally determined standard deviation in ΔC_p and ΔH_m are 4–10% (157,305) and 2–10% (306,307), respectively.

The equilibrium constant and thus free energies for unfolding can also be measured by means of a variety of indirect methods as long as the unfolding for the protein follows a two-state model (155,172,308) (some complex cases may also be considered, although analysis is more difficult). Two-state denaturation does not always exist, as has been demonstrated with staphylococcal nuclease mutants (for review, see Ref. 306). However, it has been suggested that multiple unfolded states may be considered as a single "macrostate" (309). To determine two-state folding, Shirley (310) describes an unfolding transition that should be a single step without any plateau or shoulder.

These indirect methods involve using optical techniques of various types (e.g., fluorescence and UV spectroscopies, circular dichroism, light scattering, bioassay, immunoassay, enzyme assay, chromatography, sedimentation, electrophoresis) to monitor the effects of a perturbant on protein structure. All of these methods are sensitive to three-dimensional protein structure and since, it is common for the unfolding transition to be cooperative, the transition measured by these techniques is sharp, occurring over a narrow temperature or denaturant concentration range. The ΔG for unfolding can be obtained from thermal or urea–guanidine hydrochloride denaturation curves. As can be seen in Fig. 11, the curves have linear portions at the extremes (i.e., the lowest and highest temperature or denaturant concentrations) and a sharp transition. Potential for error in experiments of these types is brought about by collection of too few points at the extremes used to extrapolate into the transition region. The extrapolated points are used to calculate the fraction of native (F_n) at each point. This is given by the equation

$$F_D = \frac{y - y_D}{y_D - y_D} \tag{21}$$

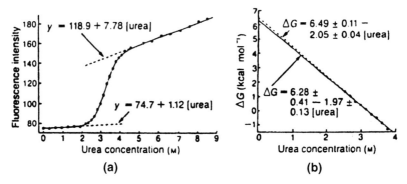

Fig. 11   (a) Urea denaturation curve for Rnase A in 30 mM formate buffer, pH 3.55, 25°C based on fluorescence measurements at 305 nm. The lines and equations are least-squares fits to the pre- and posttransition baseline. (b) ΔG calculated using Eq. 13, 21–22 as a function of urea concentration. (Reprinted with premission from Trends in Biotechnology 1990, 8, pg. 94.)

where y is the point from the experimental curve, and y_N and y_D are the values for the native and denatured states, respectively, which have been extrapolated into the transition region (see Fig. 11). The equilibrium constant can be determined from

$$K = \frac{F_n}{1 - F_n} \qquad (22)$$

and the ΔG from Eq. 13. The problem with this determination is that the ΔG obtained is in the presence of perturbant, while the desirable value is that in the absence of perturbant. The ΔG values can be extrapolated to zero perturbant concentration, but this often involves a long extrapolation and assumes that ΔG is linear with perturbant concentration. Some suggested empirical methods for extrapolating rely on attempting to sort out individual amino acid contributions to the energy to transfer an amino acid from an aqueous solvent to a denaturant solvent (311).

The advantage of these indirect techniques for determining the value of ΔG is that often much less protein is needed to perform the experiment, especially in measurements that involve the use of spectroscopy. Sample concentrations are typically 0.01 mg/mL, and a range of concentrations can be used to detect self-association. Also, solvent denaturation experiments are fast and simple. Commercially available instruments carry out automated chemical and thermal denaturation using fluorescence, CD, IR, or UV-VIS spectroscopy as a probe of protein structure (312). The disadvantage of indirect methods is that a two-state model must be followed. These techniques do not allow for the direct measure of an intermediate state. Finally, as mentioned above, a lack of data points or inaccuracies of data points in determining the baselines can significantly affect the thermodynamic parameters that are extracted.

A model-independent approach to analyze denaturant induced unfolding has been developed (185). As described earlier for indirect measurements of ΔG, the native protein is titrated with a perturbant, and an optical technique, fluorescence spectroscopy, is used to monitor the denaturation. However, rather than assuming that ΔG has a linear dependence with perturbant concentration, a nonlinear extrapolation is used. The method involves measurement of denaturant profiles at several temperatures. All parameters can be obtained from the experimental solvent denaturation data. This technique appears to show promise, but its validity needs to be further established before it can necessarily replace the linear extrapolation method described above.

Statistical mechanical theory has also been used to predict a protein's T_m as a function of pH (313). From physical properties of the constitutive amino acids, values for apomyoglobin were generated that were in agreement with experimental data from the literature. The calculations are based on the number of nonpolar groups, the protein chain length, the temperature-dependent free energy of transfer of each amino acid, and the values of pK_a as a function of temperature.

B. Long-Term Protein Stability

Accelerated stability testing is a common method for the prediction of the long-term storage conditions for pharmaceutical products (308,314). This test typically involves storing the product at elevated temperatures, above normal storage conditions, and monitoring the product's structural and/or functional integrity with time (160,315). Often, the denaturation is a first order process, that is,

$$A_t = A_0 e^{-kt} \tag{23}$$

where A_t and A_0 are the activities or concentrations at time t and time zero. From these data, the rate of degradation can be calculated at each of the different temperatures. The rates may then be fit to the Arrhenius equation:

$$\ln k = \ln \propto - \frac{\Delta E_a}{RT} \tag{24}$$

which can be used to predict rates at any temperature, most importantly, normal storage conditions (314). Arrhenius kinetics have been observed for a number of reactions involving peptides and proteins (4,15,316–322), including protein folding stability (164). When Arrhenius kinetics apply, accelerated testing provides an accurate method for the determination of product shelf life.

However, protein degradation may involve inactivation by several processes (4), and their complexity cannot be described by Arrhenius kinetics (316). Accelerated testing cannot be used to accurately predict shelf lives for protein formulations in which the loss of protein activity involves two or more consecutive simultaneous reactions that have different activation energies. In this reaction scheme, the rate of disappearance of the parent compound will not be identical with the rate of accumulation of the final product, making the plots of activity versus time complex (308). In this case, measurements of the parent compound, intermediate species, and final product must be made.

Potential errors can also arise in products that have inhomogeneous compositions, i.e. lyophilized formulations created by poor process design (308). For example, a nonuniform distribution of phases, i.e. crystalline amorphous phases can be set up in the freeze-dried cake, leading to a biphasic stability profile. This can lead to a product appearing to have a fast rate of degradation followed by a slower one at longer times (or vice versa).

Finally, accelerated stability testing can be misleading when the protein's glass transition temperature T_g, falls within the range over which the testing is being performed (308). During lyophilization, many components of the formulation are rendered amorphous. When these products are subject to accelerated stability testing, some of the samples may be stored at temperatures above their T_g, while other temperatures are below. The kinetics of reactions would not be expected to be the same in the glassy and the rubbery (i.e., viscoelastic) states. Rather, the kinetics at temperatures above T_g may follow empirical relationships such as VTF equation or WLF (Williams–Landel–Ferry) kinetics (323). WLF kinetics are described using the following equation:

$$\text{Log}(k) = -\frac{C_1(T - T_g)}{C_2 + (T - T_g)} \tag{25}$$

where C_1 and C_2 are constants related to the free volume (which would naturally play a bigger role above T_g). Thus the reaction rate correlates with $T - T_g$ rather than T (324,325). Likewise, any other phase transition (i.e., crystallization of an amorphous component), within a protein formulation that occurs in the temperature range of the stability tests would require care in applying appropriate rate equations for the prediction of product shelf life. Stability testing should be performed over a wide range in temperature to be sure that the kinetic behavior is well understood. Linear fits to Eq. 24 can be obtained even if non-Arrhenius kinetics are followed in cases of an experimental temperature range that is too small.

Accelerated stability testing can be a useful technique; however, there are many pitfalls. One should acquire a thorough understanding and characterization of the degradative mechanism(s) before attempting to interpret degradation profiles and the prediction of activity under different conditions. The FDA reminds the biologics community that accelerated testing can be used as supportive data but cannot be substituted for real-time data for product approval and labeling (315).

REFERENCES

1. D. T.-Y. Liu, Deamidation: A source of microheterogeneity in pharmaceutical proteins. Trends Biotechnol 10:364–9 (1992).
2. R. Bischoff and H. V. J. Kolbe, Deamidation of asparagine and glutamine residues in proteins and peptides: Structural determination and analytical methodology. J. Chromatog. B 662:261–278 (1994).
3. H. T. Wright, Nonenzymatic deamidation of asparaginyl and glutaminyl residues in proteins. Crit. Rev. Biochem. Mol. Biol. 26:1–52 (1991).
4. J. L. Cleland, M. F. Powell, and S. J. Shire, The development of stable protein formulations: A close look at protein aggregation, deamidation, and oxidation. Crit. Rev. Ther. Drug Carrier Syst. 10:307–377 (1993).
5. B. A. Johnson and D. W. Asward, Deamidation and isoaspartate formation during in vitro aging of purified proteins. In: D. W. Aswad, ed. Deamidation and Isoaspartate Formation in Peptides and Proteins. ed. Boca Raton, FL: CRC Press, 1995, p. 91.
6. G. Teshima, W.S. Hancoch, and E. Canova-Davis, Effect of deamidation and isoaspartate formation on the activity of proteins. In: D. W. Asward, ed. Deamidation and Isoaspartate Formation in Peptides and Proteins. Boca Raton, FL: CRC Press, 1995 p. 167.
7. A. R. Friedman, A. K. Ichhpurani, D. M. Brown, R. M. Hillman, L. F. Krabill, R. A. Martin, H. A. Zurcher-Neely, and D. M. Guido, Degradation of growth hormone releasing factor analogs in neutral aqueous solution is related to deamidation of asparagine residues. Int. J. Prot Peptide Res. 37:14–20 (1991).
8. R. W. Gracy, K. U. Yuksel, and A. Gomez-Puyou, Deamidation of triosephosphate isomerase in vitro and in vivo. In: D. W. Asward, ed. Deamidation and Isoaspartate Formation in Peptides and Proteins. Boca Raton, FL: CRC Press, 1995, p. 133.
9. T. H. Wright, Amino acid abundance and sequence data: Clues to the biological significance of non-enzymatic asparagine and glutamine deamidation in proteins. In: D. W. Asward, ed. Deamidation and Isoaspartate Formation in Peptides and Proteins. Boca Raton, FL: CRC Press, 1995. p. 229.

10. T. Geiger and S. Clarke, Deamidation, isomerization, and racemization at aspar-aginyl and aspartyl residues in peptides. J. Biol. Chem. 262:785–794 (1987).

11. R. C. Stephenson and S. Clarke, Succinimide formation from aspartyl and asparaginyl peptides as a model for the spontaneous degradation of proteins. J. Biol. Chem. 264:6164–6170 (1989).

12. R. Tyler-Cross and V. Schrich, Effects of amino acid sequence, buffers, and ionic strength on the rate and mechanism of deamidation of asparagine residues in small peptides. J. Biol. Chem. 266:22549–2256 (1991).

13. H. Tomizawa, H. Yamada, K. Wada, and T. Imoto, Stabilization of lysozyme against irreversible inactivation by suppression of chemical reactions. J. Biochem. 117:635–639 (1995).

14. M. Zhao, J. L. Bada, and T. J. Ahern, Racemization rates of asparagine-aspartic acid residues in lysozyme at 100°C as a function of pH. Bioorg. Chem. 17:36–40 (1989).

15. K. Patel and R. T. Borchardt, Chemical pathways of peptide degradation. II. Kinetics of deamidation of an asparaginyl residue in a model hexapeptide. Pharm. Res. 7:703–711 (1990).

16. T. V. Brennan and S. Clarke, Deamidation and isoaspartate formation in model synthetic peptides. The effects of sequence and solution environment. In: D. W. Asward, ed. Deamidation and Isoaspartate Formation in Peptides and Proteins. Boca Raton FL: CRC Press 1995, p. 65.

17. C. E. M. Voorten, W. A. deHaard-Hoekman, P. J. M. van den Oeelaar, H. Bloemendal, and W. W. deJong, Spontaneous peptide bond cleavage in aging α-crystalline through a succinimide intermediate. J. Biol. Chem. 263:19020–19023 (1988).

18. N. P. Bhatt, K. Patel, and R. T. Borchardt, Chemical pathways of peptide degradation. I. Deamidation of adrenocorticotropic hormone. Pharm. Res. 7:593–599 (1990).

19. R. T. Darrington and B. D. Anderson, Evidence for a common intermediate in insulin deamidation and covalent dimer formation: Effects of pH and aniline trapping in dilute acidic solutions. J. Pharm. Sci. 84:275–282 (1995).

20. A. DiDonato and G. D'Alessio, Heterogeneity of bovine seminal ribonuclease. Biochemistry 20:7232–7237 (1981).

21. R. I. Senderoff, S. C. Wootton, A. M. Boctor, T. M. Chen, A. B. Giordani, T. N. Julian, and G. W. Radebaugh, Aqueous stability of human epidermal growth factor 1–48. Pharm. Res. 11:1712–1720 (1994).

22. A. B. Robinson and C. J. Rudd, Deamidation of glutaminyl and asparginyl residues in peptides and proteins, Curr. Top. Cell. Regul. 8:247–295 (1974).

23. J. W. Scotchler and A. B. Robinson, Deamidation of glutaninyl residues: Dependence on pH, temperature, and ionic strength. Anal. Biochem. 59:319–322 (1974).

24. M. D. DiBiase and C. T. Rhodes, The design of analytical methods for use in topical epidermal growth factor product development. J. Pharm. Pharmacol. 43:553–558 (1991).

25. C. Georg-Nascimento, J. Lowenson, M. Borissenko, M. Carlderon, A. Medina-

Selby, J. Kuo, S. Clarke, and A. Randolph, Replacement of a labile aspartyl residue increases the stability of human epidermal growth factor. Biochemistry 29:9584–9591 (1990).

26. A. DeDomato, M. A. Ciardiello, M. deNigris, R. Piccoli, L. Mazzarella, and G. D'Alessio, Selective deamidation of ribonuclease A. J. Biol. Chem. 268: 4745–4751 (1993).

27. C. Aliyai, J. P. Patel, L. Carr, and R. T. Borchardt, Solid state chemical instability of an asparaginyl residue in a model hexapeptide, PDA. J. Pharm. Sci. Technol. 48:167–173 (1994).

28. J. H. McKerrow and A. B. Robinson, Deamidation of asparaginyl residues as a hazard in experimental protein and peptide procedures. Anal. Biochem. 42: 565–568 (1971).

29. S. Capasso, I. Mazzarella, and A. Zagari, Deamidation via cyclic imide of asparaginyl peptides: Dependence on salts, buffers and organic solvents. Peptide Res. 4:234–238 (1991).

30. B. A. Johnson, J. M. Shirokawa, and D. W. Aswad, Deamidation of calmodulin at neutral and alkaline pH: quantitative relationships between ammonia loss and the susceptibility of calmodulin to modification by protein carboxyl methyltransferase. Arch. Biochem. Biophys. 268:276–286 (1989).

31. H. Tomizawa, H. Yamada, K. Tanigawa, and T. Imoto, Effect of additives on irreversible inactivation of lysozyme at neutral pH and 100°C. J. Biochem. 117: 364–369 (1995).

32. R. Lura and V. Schrich, Role of peptide conformation in the rate and mechanism of deamidation of asparaginyl residues. Biochemistry 27:7671–7677 (1988).

33. T. V. Brennan and S. Clarke, Spontaneous degradation of polypeptides at aspartyl and asparaginyl residues: Effects of the solvent dielectric. Protein Sci. 2: 331–338 (1993).

34. R. C. Weast, ed. CRC Handbook of Chemistry and Physics, 54th ed., Boca Raton, FL: CRC Press, 1973.

35. Y. Nozaki and C. Tanford, The solubility of amino acids and two glycine peptides in aqueous ethanol and dioxane solutions. Establishment of a hydrophobicity scale. J. Biol. Chem. 246:2211–2217 (1971).

36. S. J. Leach and H. Lindy, The kinetics of hydrolysis of the amide group in proteins and peptides. Trans. Faraday Soc. 49:915–920 (1953).

37. K. Patel, Stability of adrenocorticotropic hormone (ACTH) and pathways of deamidation of asparginyl residue in hexapeptide segments. In: Y. J. Wang and R. Pearlman, eds. Characterization of Protein and Peptide Drugs. New York: Plenum Press, 1993, p. 201.

38. S. A. Bernhard, A. Berger, J. H. Carter, E. Katchalski, M. Seal, and Y. Shalitin, Co-operative effects of functional groups in peptides. I. Aspartyl–serine derivatives. J. Am. Chem. Soc. 84:2421–2434 (1962).

39. T. V. Brennan and S. Clarke, Effects of adjacent histidine and cysteine residues on the spontaneous degradation of asparaginyl- and aspartyl-containing peptides, Int. J. Protein Peptide Res. 45:547–553 (1995).

40. W. J. Chazin and A. K. Kossiakoff, The role of secondary and tertiary structure in intramolecular deamidation of proteins. In: D. W. Aswad, ed. Deamidation and Isoaspartate Formation in Peptides and Proteins. Boca Raton FL: CRC Press, 1995, p. 193.
41. W. E. Riha, H. V. Izzo, J. Zhang, and C.-T. Ho, Nonenzymatic deamidation of food proteins. Crit. Rev. Food Sci. Nutr. 36 225–255 (1996).
42. T. E. Creighton, Proteins: Structures and Molecular Properties. New York: W. H. Freeman, 1984.
43. A. K. Kossiakoff, Tertiary structure is a principal determinant to protein deamidation. Science 240:191–194 (1988).
44. C. Toniolo, G. M. Bonora, M. Mutter, and V. N. Rajasekharan-Pillai, The effect of the insertion of a proline residue on the solution conformation of host peptides. Makromol. Chem. 182 2007–2014 (1981).
45. R. T. Darrington and B. D. Anderson, The role of intramolecular nucleophilic catalysis and the effects of self-association on the deamidation of human insulin at low pH. Pharm. Res. 11:784–793 (1994).
46. R. T. Darrington and B. D. Anderson, Effects of insulin concentration and self-association on the partitioning of its A-21 cyclic anhydride intermediate to desamido insulin and covalent dimer. Pharm. Res. 12:1077–1084 (1995).
47. A. Q. Sun, K. U. Yuksel, and R. W. Gracy, Terminal marking of triosephosphate isomerase. Consequences of deamidation. Arch. Biochem. Biophys. 322: 361–368 (1995).
48. W. Norde, The behavior of proteins at interfaces, with special attention to the role of the structure stability of the protein molecule. Clin. Mater. 11:85–91 (1992).
49. Y. Zhang, K. U. Yuksel, and R. W. Gracy, Terminal marking of avian triosephosphate isomerase by deamidation and oxidation. Arch. Biochem. Biophys. 317:112–120 (1995).
50. H. Tomizawa, H. Yamada, and T. Imoto, The mechanism of irreversible inactivation of lysozyme at pH 4 and 100°C. Biochemistry 33:13032–13037 (1994).
51. D. B. Volkim, A. M. Verticelli, M. W. Bruner, K. E. Marfia, P. K. Tsai, M. K. Sardana, and C. R. Middaugh, Deamidation of polyanion-stabilized acidic fibroblast growth factor. J. Pharm. Sci. 84:7–11 (1995).
52. J. Janin and S. Wodak, Conformation of amino acid side-chains in proteins, J. Mol. Biol. 125:357–386 (1978).
53. J. Hamada, Deamidation of food proteins to improve functionality. Crit. Rev. Food Sci. Nutr. 34:283–292 (1994).
54. T. Wright, Sequence and structure determinants of the nonenzymatic deamidation of asparagine and glutamine residues in proteins. Protein Eng. 4:283–294 (1991).
55. P. J. Halling, Protein-stabilized foams and emulsions. Crit. Rev. Food Sci. Nutr. 15:155–203 (1981).
56. R. G. Strickley and B. D. Anderson, Solid state stability of human insulin II.

Effect of water on reaction intermediate partitioning in lyophiles from pH 2–5 solutions: Stabilization against covalent dimer formation. J. Pharm. Sci. 86: 645–653 (1997).

57. S. Li, C. Schoneich, and R. T. Borchardt, Chemical instability of protein pharmaceuticals. Mechanism of oxidation and strategies for stabilization. Biotechnol. Bioeng. 48 490–500 (1995).

58. T. H. Nguyen, Oxidation in degradation of protein pharmaceuticals. In: J. L. Cleland and R. Langer, eds. Formulation and Delivery of Proteins and Peptides, ACS Symposium 567, American Chemical Society, Washington, DC: 1994, p. 59.

59. D. M. Johnson and L. C. Gu, Autoxidation and antioxidants. In: J. Swarbrick and J. C. Boyland, eds. Encyclopedia of Pharmaceutical Technology, Vol. 1. New York: Marcel Dekker, 1994, pp. 415–449.

60. Y. C. Yang and M. A. Hanson, Parenteral formulation of proteins and peptides: Stability and stabilizers. J. Parenter. Sci. Technol. 42:s3–s26 (1988).

61. C. M. Cross, A. Z. Reznick, L. Packers, P. A. Davis, Y. J. Suzuki, and B. Halliwell, Oxidative damage to human plasma proteins by ozone. Free Radical Res. Commun. 15:347–352 (1992).

62. C. N. Oliver, B. Ahn, M. E. Wittenberger, R. L. Levine, and E. R. Stadtman, Age-related alterations of enzymes may involve mixed-function oxidation reactions. In: R. C. Adelman and E. E. Dekker, eds. Modifications of Proteins During Ageing. New York: Alan R. Liss, 1985, p. 39.

63. B. M. Babior, Oxygen-dependent microbial killing by phagocytes. New Engl. J. Med. 298 659–668 (1978).

64. J. Fransson, E. Florin-Robertsson, K. Axelsson, and C. Nyhlen, Oxidation of human insulin-like growth factor I in formulation studies: Kinetics of methionine oxidation in aqueous solution and in solid state. Pharm. Res. 13:1252–1257 (1996).

65. D. C. Cipolla and S. J. Shire, Analysis of oxidized human relaxin by reversed phase HPLC, mass spectroscopy and bioassay. In: J. J. Villafranca, ed. Techniques in protein chemistry, vol. II. New York: Academic Press, 1990 pp. 543–555.

66. E. R. Stadtman, Oxidation of free amino acids and amino acid residues in proteins by radiolysis and by metal-catalyzed reactions. Annu. Rev. Biochem. 62: 797–821 (1993).

67. W. Vogt, Oxidation of methionyl residues in proteins: Tools, targets and reversal. Free Radical Biol. Med. 18:93–105 (1995).

68. A. A. Hussain, R. Awad, P. A. Crooks, and L. W. Dittert, Chloramine-T in radiolabeling techniques. Anal. Biochem. 214:495–499 (1993).

69. V. Milajlovic, O. Cascone, and M. J. Biscoglio deJimenez Bonino, Oxidation of methionine residue in equine growth hormone by chloramine-T. Int. J. Biochem. 25:1189–1193 (1993).

70. P. Riesz and T. Kondo, Free radical formation induced by ultrasound and its biological implications. Free Radical Biol. Med. 13:247–270 (1992).

71. S. Pinamonti, M. C. Chicca, M. Muzzoli, A. Papi, L. M. Fabbri, and A. Ciaccia, Oxygen radical scavengers inhibit clastogenic activity induced by sonication of human serum. Free Radical Biol. Med. 16:363–371 (1994).

72. T. H. Nguyen, J. Burnier, and W. Meng, The kinetics of relaxin oxidation by hydrogen peroxide. Pharm. Res. 10:1563–1571 (1993).

73. S. Li, T. H. Nguyen, C. Schoneich, and R. T. Borchardt, Aggregation and precipitation of human relaxin induced by metal-catalyzed oxidation. Biochemistry 34:5762–5772 (1995).

74. K. L. Maier, E. Matejkova, H. Hinze, L. Leuschel, H. Weber, and I. Beck-Speier, Different selectivities of oxidants during oxidation of methionine residues in the α_1-proteinase inhibitor, FEBS Lett. 250:221–256 (1989).

75. C. Schoneich, F. Zhao, G. S. Wilson, and R. T. Borchardt, Iron-thiolate induced oxidation of methionine to methionine sulfoxide in small model peptides. Intramolecular catalysis by histidine. Biochim. Biophys. Acta 1168:307–322 (1993).

76. S. Li, C. Schoneich, and R. T. Borchardt, Chemical pathways of peptide degradation. VIII. Oxidation of methionine in small model peptides by prooxidant/transition metal ion systems: Influence of selective scavengers for reactive oxygen intermediates. Pharm. Res. 12:348–355 (1995).

77. B. Halliwell, Superoxide-dependent formation of hydroxyl radicals in the presence of iron chelates. FEBS Lett. 92:321–326 (1978).

78. B. Halliwell and J. M. C. Gutteridge, Formation of a thiobarbituric-acid-reactive substance from deoxyribose in the presence of iron salts. FEBS Lett. 128: 347–352 (1981).

79. J. M. C. Gutteridge, R. Richmond, and B. Halliwell, Oxygen free-radicals in lipid peroxidation: Inhibition by the protein caeruloplasmin. FEBS Lett. 112 269–272 (1980).

80. E. R. Stadtman, Metal ion-catalyzed oxidation of proteins: Biochemical mechanisms and biological consequences. Free Radical Biol. Med. 9:315–325 (1990).

81. L.-C. The, L. J. Murphy, N. L. Huq, A. S. Surus, H. G. Friesen, L. Lazarus, and G. E. Chapman, Methionine oxidation in human growth hormone and human chorionic somatomammotropin. J. Biol. Chem. 262:6472–6477 (1987).

82. S. E. Ealick, W. J. Cook, S. Vijay-Kumar, M. Carson, T. L. Nagabhushan, P. P. Trotta, and C. E. Bugg, Three-dimensional structure of recombinant human interferon. Science 252:698–702 (1991).

83. R. G. Keck, The use of t-butyl hydroperoxide as a probe for methionine oxidation in proteins. Anal. Biochem. 236:56–62 (1996).

84. N. P. Neumann, Oxidation with hydrogen peroxide. Methods Enzymol. 25: 393–400 (1972).

85. W. K. Paik and S. Kim, Protein methylation: Chemical, enzymological, and biological significance. Adv. Enzymol. 42:227–286 (1975).

86. P. K. Sysak, C. S. Foote, and T.-Y. Ching, Chemistry of singlet oxygen. XXV. Photooxygenation of methionine. Photochem. Photobiol. 26:19–27 (1977).

87. C. Schoneich, A. Aced, and K.-D. Asmus, Halogenated peroxyl radicals as

two-electron-transfer agents. Oxidation of organic sulfides to sulfoxides. J. Am. Chem. Soc. 113:375–376 (1991).

88. D. T. Sawyer, C. Kang, A. Llobet, and C. Redman, Fenton Reagents (1:1 Fe(II)L$_x$/HOOH) React via [L$_x$Fe(II)OOH(BH$^+$)] (1) as hydroxylases (RH $\Rightarrow$ ROH), not as generators of free hydroxyl radicals (HO·), J. Am. Chem. Soc. 115:5817–5818 (1993).

89. I. Yamajaka and L. H. Piette, EPR spin-trapping study on the oxidizing species formed in the reaction of the ferrous ion with hydrogen peroxide. J. Am. Chem. Soc. 113:7588–7593 (1991).

90. S. Li, C. Schoneich, G. S. Wilson, and R. T. Borchardt, Chemical pathways of peptide degradation. V. Ascorbic acid promotes rather than inhibits the oxidation of methionine to methionine sulfoxide in small model peptides. Pharm. Res. 10:1572–1579 (1993).

91. N. E. Good, G. D. Winget, W. Winter, T. N. Connolly, S. Izawa, and R. M. M. Singh, Hydrogen ion buffers for biological research. Biochemistry 5:467–477 (1966).

92. M. Hicks and J. M. Gebicki, Rate constants for reaction of hydroxy radicals with tris, tricine and hepes buffer. FEBS Lett. 199:92–94 (1986).

93. S. Miskoski and N. A. Garcia, Influence of the peptide bond on the singlet molecular oxygen-mediated (O$_2$[$^1\Delta_g$]) photooxidation of histidine and methionine dipeptides. A kinetic study. Photochem. Photobiol. 57:447–452 (1993).

94. M. Linetsky and B. J. Ortwerth, Quantitation of the reactive oxygen species generated by the UVA irradiation of ascorbic acid–glycated lens proteins. Photochem. Photobiol. 63:649–655 (1996).

95. T. Funakoshi, M. Abe, M. Sakata, S. Shoji, and Y. Kubota, The functional side of placental anticoagulant protein: Essential histidine residue of placental anticoagulant protein. Biochem. Biophys. Res. Commun. 168:125–134 (1990).

96. J. Stuart, I. N. Pessah, T. G. Favero, and J. J. Abramson, Photooxidation of skeletal muscle sarcoplasmic reticulum induces rapid calcium release. Arch. Biochem. Biophys. 292:512–521 (1992).

97. M. Tomita, M. Irie, and T. Ukita, Sensitized photooxidation of histidine and its derivatives. Products and mechanism of the reaction. Biochem. 8:5149–5160 (1969).

98. P. P. Batra and G. Skinner, A kinetic study of the photochemcial inactivation of adenylate kinases of *Mycobacterium marinum* and bovine heart mitochondria. Biochim. Biophys. Acta 1038:52–59 (1990).

99. C. Giulive, M. Sarcansky, E. Rosenfeld, and A. Boveris, The photodynamic effect of rose bengal on proteins of the mitochondrial inner membrane. Photochem. Photobiol. 52:745–751 (1990).

100. B. J. Ortwerth, M. Linetsky, and P. R. Olesen, Ascorbic acid glycation of lens proteins produces UVA sensitizers similar to those in human lens. Photochem. Photobiol. 62:454–462 (1995).

101. A. Amici, R. L. Levine, L. Tsai, and E. R. Stadtman, Conversion of amino acid

residues in proteins and amino acids homopolymers to carbonyl derivatives by metal-catalyzed oxidation reactions. J. Biol. Chem. 264:3341–3346 (1989).

102. R.-Z. Cheng and S. Kawakishi, Site-specific oxidation of histidine residues in glycated insulin mediated by Cu^{+2}. Eur. J. Biochem. 223:759–764 (1994).

103. J.-I. Ueda, T. Ozawa, M. Miyazaki, and Y. Fujiwara, Activation of hydrogen peroxide by copper(II) complexes with some histidine-containing peptides and their SOD-like activities. J. Inorganic Biochem. 55:123–130 (1994).

104. K. Uchida and S. Kawakishi, Identification of oxidized histidine generated at the active site of Cu, Zn-superoxide dismutase exposed to H_2O_2. J. Biol. Chem. 269:2405–2410 (1994).

105. S. A. Lewisch and R. L. Levine, Determination of 2-oxohistidine by amino acid analysis. Anal. Biochem. 231:440–446 (1995).

106. K. A. Engelka and T. Maciaj, Inactivation of human fibroblast growth factor-1 (FGF-1) activity by interaction with copper ions involves FGF-1 dimer formation induced by copper-catalyzed oxidation. J. Biol. Chem. 267:11307–11315 (1992).

107. R. Munday, Toxicity of thiols and disulphides: Involvement of free-radical species. Free Radical Biol. Med. 7:659–673 (1989).

108. P. K. Tsai, D. B. Volkin, J. M. Dabora, K. C. Thompson, M. W. Bruner. J. O. Gress, B. Matuszewska, M. Keogan, J. V. Bondi, and C. R. Middaugh, Formulation design of acidic fibroblast growth factor. Pharm. Res. 10:649–659 (1993).

109. E. H. Muslin, D. Li, F. J. Stevens, M. Donnelly, M. Schiffer, and L. E. Anderson, Engineering a domain-locking disulfide into a bacterial malate dehydrogenase produces a redox-sensitive enzymer. Biophys. J. 68:2218–2223 (1995).

110. C. Little and P. J. O'Brien, Mechanism of peroxide-inactivation of the sulphydryl enzyme glyceraldehyde-3-phosphate dehydrogenase. Eur. J. Biochem. 10: 533–538 (1969).

111. K. Itakura, K. Uchida, and S. Kawakishi, A novel tryptophan dioxygenation by superoxide. Tetrahedron Lett. 33:2567–2570 (1992).

112. M. Nakagawa, H. Watanabe, S. Kodato, H. Okajima, T. Hino, J. L. Flippen, and B. Witkop, A valid model for the mechanism of oxidation of tryptophan to formylkynurenine—25 years later. Proc. Natl. Acad. Sci. USA 74:4730–4733 (1977).

113. Z. Maskos, J. D. Rush, and W. H. Koppenol, The hydroxylation of tryptophan, Arch. Biochem. Biophys. 296:514–520 (1992).

114. G. Kell and H. Steinhart, Oxidation of tryptophan by H_2O_2 in model systems. J. Food Sci. 55:1120–1123 (1990).

115. K. Itakura, K. Uchida, and S. Kawakishi, Selective formation of oxindole- and formylkynurenine-type products from tryptophan and its peptides treated with a superoxide-generating system in the presence of iron(III)-EDTA: A possible involvement with iron-oxygen complex. Chem. Res. Toxicol. 7:185–190 (1994).

116. W. E. Savige, Isolation and identification of some photo-oxidation products of tryptophan. Aust. J. Chem. 24:1285–1293 (1975).

117. L. A. Holt, B. Milligan, D. E. Rivett, and F. H. C. Stewart, The photodecomposition of tryptophan peptides. Biochim. Biophys. Acta 499:131–138 (1977).

118. J. D. Tassin and R.F. Borkman, The photolysis rates of some di- and tripeptides of tryptophan. Photochem. Photobiol. 32:577–585 (1980).

119. H. Templer and P. J. Thistlewaite, Flash photolysis of aqueous tryptophan, alanyl tryptophan and tryptophyl alanine. Photochem. Photobiol. 23:79–85 (1976).

120. D. D. Busath and R. C. Waldbillig, Photolysis of gramicidin A channels in lipid bilayers. Biochim. Biophys. Acta 736:28–38 (1983).

121. S. McKim and J. F. Hinton, Direct observation of differential UV photolytic degradation among the tryptophan residues of gramicidin A in sodium dodecyl sulfate micelles. Biochim. Biophys. Acta 1153:315–321 (1993).

122. C. Pigault and D. Gerard, Influence of the location of tryptophanyl residues in proteins on their photosensitivity. Photochem. Photobiol. 40:291–296 (1984).

123. T. G. Huggins, M. C. Wells-Knecht, N. A. Detorie, J. W. Baynes, and S. R. Thorpe, Formation of o-tyrosine and dityrosine in proteins during radiolytic and metal-catalyzed oxidation. J. Biol. Chem. 268:12341–12347 (1993).

124. E. R. Stadtman, Role of oxidized amino acids in protein breakdown and stability. Methods Enzymol. 258:379–393 (1995).

125. J. W. Heinecke, W. Li, H. L. Daehnke III, and J. A. Goldsten, Dityrosine, a specific marker of oxidation, is synthesized by the myeloperoxidase-hydrogen peroxide system of human neutrophils and macrophages. J. Biol. Chem. 268: 4069–4077 (1993).

126. K. Uchida, Y. Kato, and S. Kawakishi, A novel mechanism for oxidative cleavage of prolyl peptides induced by the hydroxyl radical. Biochem. Biophys. Res. Commun. 169:265–271 (1990).

127. Y. Kato, K. Uchida, and S. Kawakishi, Oxidative fragmentation of collagen and prolyl peptide by Cu(II)/H_2O_2. J. Biol. Chem. 267:23646–23651 (1992).

128. W. Garrison, Reaction mechanisms in the radiolysis of peptides, polypeptides, and proteins. Chem. Rev. 87:381–398 (1987).

129. A. Nishinaga, T. Shimizu, and T. Matsuura, Base-catalyzed oxygenation of tert-butylated phenols. J. Org. Chem. 44:2983–2988 (1979).

130. J. W. McGinity, J. A. Hill, and A. L. LaVia, Influence of peroxide impurities in polyethylene glycols on drug stability. J. Pharm. Sci. 64:356–357 (1975).

131. D. M. Johnson and W. F. Taylor, Degradation of fenprostalene in polyethylene glycol 400. J. Pharm. Sci. 73:1414–1417 (1984).

132. L. Chafetz, W.-H. Hong, D. C. Tsilifonis, A. K. Taylor, and J. Philip, Decrease in the rate of capsule dissolution due to formaldehyde from polysorbate 80 autoxidation. J. Pharm. Sci. 73:1186–1187 (1984).

133. P. Labrude, B. Chaillo, F. Bonneaux, and C. Vigneron, Freeze-drying of haemoglobin in the presence of carbohydrates. J. Pharm. Pharmacol. 32:588–589 (1980).

134. D. A. Rowley and B. Halliwell, Superoxide-dependent and ascorbate-dependent formation of hydroxyl radicals in the presence of copper salts: A physiologically significant reaction? Arch. Biochem. Biophys. 225:279–284 (1983).
135. J. M. Braughler, L. A. Duncan, and R. L. Chase, The involvement of iron in lipid peroxidation. J. Biol. Chem. 261:10282–10289 (1986).
136. S. Li, T. W. Patapoff, T. H. Nguyen, and R. T. Borchardt, Inhibitory effect of sugars and polyols on the metal-cataylzed oxidation of human relaxin. J. Pharm. Sci. 85:868–875 (1996).
137. S. J. Angyal, Complexing of carbohydrates with copper ions: A reappraisal. Carbohydr. Res. 200:181–188 (1990).
138. E. Franzini, H. Sellak, J. Hakim, and C. Pasquier, Comparative sugar degradation by (OH) produced by the iron-driven fenton reaction and gamma radiolysis. Arch. Biochem. Biophys. 309:261–265 (1994).
139. S. P. Wolff and R. T. Dean, Glucose autoxidation and protein modification. Biochem. J. 245:243–250 (1987).
140. J. V. Hunt, R. T. Dean, and S. P. Wolff, Hydroxyl radical production and autoxidative glycosylation. Biochem. J. 256:205–212 (1988).
141. P. K. Hunt and R. C. Roberts, Methionine oxidation and inactivation of α_1-Proteinase inhibitor by Cu^{+2} and glucose. Biochim. Biophys. Acta 1121:325–330 (1988).
142. S. Fujimoto, T. Nakagawa, S. Ishimitsu, and A. Ohara, On the mechanism of inactivation of papain by bisulfite. Chem. Pharm. Bull. 31:992–1000 (1983).
143. M. J. Fryer, Evidence for the photoprotective effects of vitamin E. Photochem. Photobiol. 58:304–312 (1993).
144. M. J. Pikal, K. Dellman, and M. L. Roy, Formulation and stability of freeze-dried proteins: Effects of moisture and oxygen on the stability of freeze-dried formulations of human growth hormone. Dev. Biol. Stand. 74:21–38 (1992).
145. M. J. Pikal, K. M. Dellerman, M. J. Roy, and R. M. Riggin, The effects of formulation variables on the stability of freeze-dried human growth hormone. Pharm. Res. 8:427–436 (1991).
146. R. P. Enever, A. Li Wan Po, and E. Shotton, Factors influencing decomposition rate of amitriptyline hydrochloride in aqueous solution. J. Pharm. Sci. 66:1087–1089 (1977).
147. C. Town, Moisture content in proteins: Its effects and measurement. J. Chromatogr. A. 705:115–127 (1995).
148. M. J. Hageman, The role of moisture in protein stability. Drug. Dev. Ind. Pharm. 14:2047–2070 (1988).
149. D. E. Anderson, W. J. Becktel, and F. W. Dahlquist, pH-Induced denaturation of proteins: A single salt bridge contributes 3–5 kcal/mol to the free energy of folding of T4 lysozyme. Biochemistry 29:2403–2408 (1990).
150. C. N. Pace, The stability of globular proteins. CRC Crit. Rev. Biochem. 3:1–43 (1975).
151. K. Aune and C. Tanford, Thermodynamics of the denaturation of lysozyme

by guanidine hydrochloride. II. Dependence on denaturant concentration at 25 degrees. Biochemistry 8:4586–4590 (1969).

152. P. Privalov and N. Khechinashvilli, A thermodynamic approach to the problem of stabilization of globular protein structure: A carlorimetric study. J. Mol. Biol. 86:665–684 (1974).

153. B. Madan and K. J. Sharp, Heat capacity changes accompanying hydrophobic and ionic solvation: A Monte Carlo and random network model study. J. Phys. Chem. 100:7713–7721 (1996).

154. J. T. Edsall, J. Am. Chem. Soc. 57:1506 (1935).

155. A. D. Robertson and K. P. Murphy, Protein structure and the energetics of protein stability. Chem. Rev. 97:1251–1267 (1997).

156. R. Lumry, R. Biltonen, and J. Brandts, Validity of the "two-state" hypothesis for conformational transitions of proteins. Biopolymers 4:917–944 (1966).

157. C. N. Pace and D. V. Laurents, A new method for determining the heat capacity change for protein folding. Biochemistry 28:2520–2525 (1989).

158. J. A. schellman, The thermodynamic stability of proteins. Annu. Rev. Biophys. Biophys Chem. 16:115–137 (1987).

159. B.-L. Chen and T. Arakawa, Stabilization of recombinant human keratinocyte growth factor by osmolytes and salts. J. Pharm. Sci. 85:419–422 (1996).

160. C. Fágáin, Understanding and increasing protein stability. Biochim. Biophys. Acta. 1252:1–14 (1995).

161. D. A. Parsell and R. T. Sauer, The structural stability of a protein is an important determinant of its proteolytic susceptibility in *Escherichia coli*. J. Biol. Chem. 264:7590–7595 (1989).

162. P. L. Privalov and S. J. Gill, Stability of protein structure and hydrophobic interaction. Adv. Protein Sci. 39:191–234 (1988).

163. K. A. Dill and D. O. V. Alonsa, In: E. L. Winnacker and R. Huber, eds. Protein Structure and Protein Engineering. Berlin, Springer-Verlag, 1988, p. 51.

164. M. L. Scalley and D. Baker, Protein folding kinetics exhibit an Arrhenius temperature dependence when corrected for the temperature dependence of protein stability. Proc. Natl. Acad. Sci. U.S.A. 94:10636–10640 (1997).

165. A. Dong, S. J. Prestrelski, S. D. Allison, and J. F. Carpenter, Infrared spectroscopic studies of lyophilization- and temperature-induced protein aggregation. J. Pharm. Sci. 84:415–424 (1995).

166. C. Tanford, R. M. Pain, and N. S. Otchin, Unfolding of hen egg white lysozyme by guanidine hydrochloride. J. Mol. Biol. 15:489 (1966).

167. P. L. Privalov, N. N. Khechinashvili, and B. P. Atanasov, Thermodynamic analysis of thermal transitions in globular proteins. I. Calorimetric study of ribotrypsinogen, ribonuclease and myoglobin. Biopolymers 10:1865 (1971).

168. J. A. Schellman, Selective binding and solvent denaturation. Biopolymers 26:549–559 (1987).

169. C. N. Pace and D. V. Laurents, RNases A and T1: pH dependence and folding. Biochemistry 29:2520–2525 (1990).

170. C. Q. Hu, J. M. Sturtevant, J. A. Thomson, R. E. Erickson, and C. N. Pace, Thermodynamics of ribonuclease T1 denaturation. Biochemistry 31:4876–4882 (1992).

171. C. N. Pace, Conformational stability of globular proteins. Trends Biol. Sci. 15: 14–17 (1990).

172. C. N. Pace, Measuring and increasing protein stability. Trends Biotechnol. 8: 93–98 (1990).

173. G. Feller and C. Gerday, Psychrophilic enzymes: Molecular basis of cold adaptation. Cell. Mol. Life Sci. 53:830–841 (1997).

174. R. Jaenicke, Protein stability and protein folding. CIBA Found. Symp. 161: 206–216 (1991).

175. R. Jaenicke, Protein stability and molecular adaptation to exteme conditions. Eur. J. Biochem. 202:715–728 (1991).

176. G. N. Somero, Temperature and proteins: Little things can mean a lot. News Physiol. Sci. 11:72–77 (1996).

177. D. A. Cowan, Protein stability at high temperatures. Essays Biochem. 29:193–207 (1995).

178. D. B. Volkin, C. R. Middaugh, The effect of temperature on protein structure. In T. J. Ahern, and M. C. Manning, eds. Stability of Protein Pharmaceuticals, Part A: Chemical and Physical Pathways of Protein Degradation. New York: Plenum Press, 1992, pp. 215–247.

179. Y. Suzuki, Proc. Jpn. Acad. Sci. Ser. B:146–148 (1990).

180. M. F. Perutz and H. Raidt, Stereochemical basis of heat stability in bacterial ferredoxins and in haemoglobin A2. Nature 255:256–259 (1975).

181. B. A. Shirley, Protein conformational stability estimated from urea, guanidine hydrochloride, and thermal denaturation curves. In: T. J. Ahern and M. C. Manning, eds. Stability of Protein Pharmaceuticals, Part A: Chemical and Physical Pathways of Protein Degradation. New York: Plenum Press, 1992, pp. 167–194.

182. W. Pfeil, U. Gesierich, G. R. Kleemann, and R. Sterner, Ferredoxin from the hyperthermophile *Thermotoga maritima* is stable beyond the boiling point of water. J. Mol. Biol. 272:591–596 (1997).

183. R. L. Baldwin, Temperature dependence of the hydrophobic interaction in protein folding. Proc. Natl. Acad. Sci. U.S.A. 83:8069–8072 (1986).

184. P. L. Privalov, Stability of proteins: Small globular proteins. Adv. Protein Chem. 33:167–241 (1979).

185. B. Ibarra-Molero and J. M. Sanchez-Ruiz, A model independent, nonlinear extrapolation procedure for the characterization of protein folding energetics from solvent-denaturation data. Biochemistry 35:14689–14702 (1996).

186. V. Rishi, F. Anjum, F. Ahmad, and W. Pfeil, Role of non-compatible osmolytes in the stabilization of proteins during heat stress. Biochem. J. 329:137–143 (1998).

187. S. J. Prestrelski, T. Arakawa, and J. F. Carpenter, Separation of freezing- and drying-induced denaturation of lyophilized proteins using stress-specific stabilization. Arch. Biochem. Biophys. 303:465–473 (1993).

188. G. N. Somero, Proteins and temperature. Annu. Rev. Physiol. 57:43–68 (1995).
189. T. Hottiger, T. Boller, and A. Wiemken, Rapid change of heat and desiccation tolerance correlated with changes of trehalsoe content in *Saccharomyces cerevisiae* cells subjected to temperature shifts. FEBS Lett. 22:113–115 (1987).
190. M. J. Akers, N. Milton, S. R. Byrn, and S. L. Nail, Glycine crystallization during freezing: The effects of salt form, pH, and ionic strength. Pharm. Res. 12:1457–1461 (1995).
191. K. Izutsu, S. Yoshioka, and Y. Takeda, Protein denaturation in dosage forms measured by differential scanning calorimetry. Chem. Pharm. Bull. 38:800–803 (1990).
192. S. Yoshioka, Y. Aso, Y. Nakai, and S. Kojima, Effect of high molecular mobility of poly(vinyl alchohol) on protein stability of lyophilized γ-globulin formulations. J. Pharm. Sci. 87:147–151 (1998).
193. S. Yoshioka, Y. Aso, and S. Kojima, Dependence of the molecular mobility and protein stability of freeze-dried γ-globulin formulations on the molecular weight of dextran. Pharm. Res. 14:736–741 (1997).
194. S. J. Prestrelski, K. A. Pikal, and T. Arakawa, Optimizations of lyophilization conditions for recombinant human interleukin-2 by dried-state conformational analysis using Fourier-transform infrared spectroscopy. Pharm. Res. 12:1250–1259 (1995).
195. L. N. Bell, M. J. Hageman, and L. M. Muraoka, Thermally induced denaturation of lyophilized bovine somatotropin and lysozyme as impacted by moisture and excipients. J. Pharm. Sci. 84:707–712 (1995).
196. C. A. Oksanen and G. Zografi, Molecular mobility in mixtures of absorbed water and solid poly(vinylpyrrolidone). Pharm. Res. 10:791–799 (1993).
197. B. C. Hancock, S. L. Shamblin, and G. Zografi, Molecular mobility of amorphous pharmaceutical solids below their glass transition temperatures. Pharm. Res. 12:799–806 (1995).
198. B. S. Chang, R. M. Beauvais, A. Dong, and J. F. Carpenter, Physical factors affecting the storage stability of freeze-dried interleukin-1 receptor antagonist: Glass transition and protein conformation. Arch. Biochem. Biophys. 331:249–258 (1996).
199. Y.-J. Tan, M. Oliveberg, B. Davis, and A. R. Fersht, Perturbed pK_a-values in the denatured state of proteins. J. Mol. Biol. 254:980–992 (1995).
200. A.-S. Yang and B. Honig, Structural origins of pH and ionic strength effects on protein stability. J. Mol. Biol. 237:602–614 (1994).
201. W. E. Stites, A. G. Gittis, E. Lattman, and D. Shortle, In a staphylococcal nuclease mutant the side-chain of a lysine replacing valine 66 is fully buried in the hydrophobic core. J. Mol. Biol. 221:7–14 (1991).
202. A. Rashin and B. Honig, On the environment of ionizable groups in globular proteins. J. Mol. Biol. 173:515–521 (1984).
203. M. Perutz, Mechansim of denaturation of hemoglobin by alkali. Nature 247:341–344 (1974).
204. K. Langsetmo, J. A. Fuchs, C. Woodward, and K. A. Sharp, Linkage of thiore-

doxin stability to titration of ionizable groups with perturbed pK_a. Biochemistry 30:7609–7614 (1991).

205. A. S. Yang and B. Honig, Electrostatic effects on protein stability. Curr. Opin. Struct. Biol. 2:40–45 (1992).

206. C. N. Pace, D. V. Laurents, and R. E. Erickson, Urea denaturation of barnase: pH dependence and characterization of the unfolded state. Biochemistry 31: 2728–2734 (1992).

207. C. N. Pace, D. V. Laurents, and J. A. Thomsom, pH dependence of the urea and guanidine hydrochloride denaturation of ribonuclease A and ribonuclease T1. Biochemistry 29:2564–2572 (1990).

208. T. E. Creighton, Proteins: Structures and Molecular Properties. New York: W. H. Freeman, 1993, pp. 1–48.

209. F. Inagaki, T. Kawano, I. Shimada, K. Takahashi, and T. Miyazawa, Nuclear magnetic resonance study on the microenvironments of histidine residues of ribonuclease T1 and carboxymethylated ribonuclease T1. J. Biochem. Tokyo 89:1185–1195 (1981).

210. J. Gao, M. Mrksich, F. A. Gomex, and G. M. Whitesides, Using capillary electrophoresis to follow the acetylation of the amino groups of insulin and to estimate their basicities. Anal. Chem. 67:3093–3100 (1995).

211. A.-S. Yang, and B. Honig, On the pH dependence of protein stability. J. Mol. Biol. 231:459–474 (1993).

212. M. Schaefer, M. Sommer, and M. Karplus, pH-dependence of protein stability: Absolute electrostatic free energy differences between conformations. J. Phys. Chem. 101:1663–1683 (1997).

213. J. Antosiewicz, J. A. McCammon, and M. K. Gilson, Prediction of pH-dependent properties of proteins. J. Mol. Biol. 238:415–436 (1994).

214. T. Arakawa, Y. Kita, and J. F. Carpenter, Protein–solvent interactions in pharmaceutical formulations. Pharm. Res. 8:285–291 (1991).

215. S. N. Timasheff, The control of protein stability and association by weak interactions with water: How do solvents affect these processes? Annu. Rev. Biophys. Biomol. Struct. 22:67–97 (1993).

216. F. Franks, Protein hydration. In: F. Franks, ed. Protein Biotechnology. Cambridge: Parfa Ltd., 1993, pp. 395–436.

217. K. D. Collins and M. W. Washabaugh, The Hofmeister effect and the behavior of water at interfaces. Q. Rev. Biophys. 18:323–422 (1985).

218. R. L. Baldwin, How Hofmeister ion interactions affect protein stability. Biophys. J. 71:2056–2063 (1996).

219. P. H. Yancey, M. E. Clark, S. C. Hand, R. D. Bowlus, and G. N. Somero, Living with water stresses: Evolution of osmolyte systems, Science 217:1214 (1982).

220. G. Xie and S. N. Timasheff, Mechanism of the stabilization of ribonuclease A by sorbitol: Preferential hydration is greater for the denatured than for the native protein. Protein Sci. 6:211–221 (1997).

221. G. Xie and S. N. Timasheff, The thermodynamic mechanism of protein stabilization by trehalose. Biophys. Chem. 64:25–43 (1997).

222. K. Gekko and S. N. Timasheff, Mechanism of protein stabilization by glycerol: Preferential hydration in glycerol–water mixtures. Biochemistry 20: 4667–4676 (1981).

223. J. C. Lee and S. N. Timasheff, The stabilization of proteins by sucrose. J. Biol. Chem. 256:7193–7201 (1981).

224. S. N. Timasheff and T. Arakawa, Stabilization of protein structure by solvents. In: T. E. Creighton, ed *Protein Structure: A Practical Approach.* Oxford: IRL Press, 1989, p. 331.

225. T. Y. Lin and S. N. Timasheff, On the role of surface tension in the stabilization of globular proteins. Protein Sci. 5:372–381 (1996).

226. F. R. Cioci, R. Lavecchia, and L. Marrelli, Perturbation of surface tension of water by polyhydric additives: Effect on glucose oxidase stability. Biocatalysis 10:137 (1994).

227. I. Baskakov and D. W. Bolen, Forcing thermodynamically unfolded proteins to fold. J. Biol. Chem 273:4831–4834 (1998).

228. S. N. Timasheff and T. Arakawa, In: T. E. Creighton, ed., Protein Function: A practical approach. Oxford: IRL Press, 1988.

229. J. A. Schellman, Selective binding and solvent denaturation. Biopolymers 26: 549 (1987).

230. T. Arakawa, R. Bhat, and S. N. Timasheff, Why preferential hydration does not always stabilize the native structure of globular proteins. Biochemistry 29: 1924–1931 (1990).

231. J. C. Lee and S. N. Timasheff, The calculation of partial specific volumes of proteins in guanidine hydrochloride. Arch. Biochem. Biophys. 165:268–273 (1974).

232. C. N. Pace and G. R. Grimsley, Ribonuclease T1 is stabilized by cation and anion binding. Biochemistry 27:3242–3246 (1998).

233. D. Gillis and M. Rooman, Predicting protein stability changes upon mutation using database-derived potentials: Solvent accessibility determines the importance of local versus non-local interactions along the sequence. J. Mol. Biol. 272:276–290 (1997).

234. C. Tanford, J. Am. Chem. Soc. 86:2050–2059 (1964).

235. C. Tanford, Protein denaturation. C. Theoretical models for the mechanism of denaturation. Adv. Protein Chem. 24:1–95 (1970).

236. E. P. Pittz and S. N. Timasheff, Interaction of ribonuclease A with aqueous 2-methyl-2,4-pentanediol at pH 5.8. Biochemistry 17:615–623 (1978).

237. E. P. Pittz and J. Bello, Studies on bovine pancreatic ribonuclease A and model compounds in aqueous 2-methyl-2,4-pentanediol. I. Amino acid solubility, thermal reversibility of ribonuclease A, and preferential hydration of ribonuclease A crystals. Arch. Biochem. Biophys. 146:513–524 (1971).

238. B. Jirgensons, Factors determining the reconstructive denaturation of proteins in sodium dodecyl sulfate solutions. Further circular dichroism studies on structural reorganization of seven proteins. J. Protein Chem. 1:71–84 (1982).

239. J. Steinhardt, J. R. Scott, and K. S. Birdi, Differences in the solubilizing ef-

fectivness of the sodium dodecyl sulfate complexes of various proteins. Biochemistry 16:718–725 (1977).

240. E. Mori and B. Jirgensons, Effect of long-chain alkyl sulfate binding on circular dichroism and conformation of soybean trypsin inhibitor. Biochemistry 20: 1630–1634 (1981).

241. B. Jirgensons, Circular dichroism study on structural reorganization of lectins by sodium dodecyl suflate. Biochim. Biophys. Acta 623:69–76 (1980).

242. S. Tandon and P. M. Horowitz, Detergent-assisted refolding of guanidine chloride–denatured rhodanese. J. Biol. Chem. 262:4486–4491 (1987).

243. Z. J. Twardowski, K. D. Nolph, T. J. McGray, and H. L. Moore, Nature of insulin binding to plastic bags. Am. J. Hosp. Pharm. 40:579–581 (1983).

244. J. L. Bohnert and T. A. Horbett, Changes in adsorbed fibrinogen and albumin interactions with polymers indicated by decreases in detergent elutability. J. Colloid Interface Sci. 111:363–377 (1986).

245. A. S. Chawala, I. Hinberg, P. Blais, and D. Johnson, Aggregation of insulin, containing surfactants, in contact with different materials. Diabetes 34:420–424 (1985).

246. W. D. Loughheed, A. M. Albisser, H. M. Martindale, J. C. Chow, and J. R. Clement, Physical stability of insulin formulations. Diabetes 32:424–432 (1983).

247. J. Piatigorsky, J. Horowitz, and T. Simpson, Partial dissociation and renaturation of embryonic Chick d-crystalline. Characterization by ultracentrifugation and circular dichroism. Biochim. Biophys. Acta 490:279–289 (1977).

248. M. J. Schwuger and F. G. Bartnik, In: C. Gloxhuber, ed. *Anionic Surfactants*, Vol. 10. New York: Marcel Dekker, 1980, pp. 1–52.

249. J. Greener, B. A. Constable, and M. M. Bale, Interaction of anionic surfactants with gelatin: Viscosity effects. Macromolecules 20:2490–2498 (1987).

250. S. Makino, Interaction of proteins with amphiphilic substances. In: M. Kotani, ed. Advances in Biophysics, Vol. 12. Baltimore: University Park Press, 1979, pp. 131–184.

251. H. M. Rendall, Use of a surfactant selective electrode in the measurement of the binding of anionic surfactants to bovine serum albumin. J. Chem. Soc. Faraday Trans. 1:481–484 (1976).

252. G. C. Kreschek and I. Constandinidis, Ion-selective electrodes for octyl and decyl sulfate surfactants. Anal. Chem. 56:152–156 (1984).

253. K. Hayakawa, A. L. Ayub, and J. C. T. Kwak, The application of surfactant selective electrodes to the study of surfactant adsorption in colloidal suspensions. Colloids Surf. 4:389–397 (1982).

254. N. B. Bam, T. W. Randolf, and J. L. Cleland, Stability of protein formulations: Investigation of surfactant effects by a novel EPR spectroscopic technique. Pharm. Res. 12:2–11 (1995).

255. A. Lustig, A. Engel, and M. Zulauf, Density determination by analytical ultracentrifugation in a rapid dynamical gradient: Application to lipid and deter-

gent aggregates containing proteins. Biochem. Biophys. Acta 1115:89–95 (1991).

256. M. L. Smith and N. Muller, Fluorine chemical shifts in complexes of sodium trifluoroalkylsulfates with reduced proteins. Biochem. Biophys. Res. Commun. 62:723–728 (1975).

257. K. Tsuji and T. Takagi, Proton magnetic resonance, J. Biochem. 77:511–519 (1975).

258. S. Sato, C. D. Ebert, and S. W. Kim, Prevention of insulin self-association and surface adsorption. J. Pharm. Sci. 72:228–232 (1983).

259. J. R. Brennan and W. G. Gebhart, Pump-induced insulin aggregation. A problem with the biostatror. Diabetes 34:353–359 (1985).

260. S. E. Charm and B. L. Wong, Enzyme inactivation with shearing. Biotechnol. Bioeng. 12:1103–1109 (1970).

261. S. E. Charm and B. L. Wong, Shear inactivation of fibrinogen in the circulation. Science 170:466–468 (1975).

262. S. E. Charm and B. L. Wong, Shear inactivation of heparin. Biorheology 12: 275–278 (1975).

263. D. J. Burgess, J. K. Yoon, and N. O. Sahin, A novel method of determination of protein stability. J. Parenter. Sci. Technol. 46:150–155 (1992).

264. S. J. Prestrelski, N. Tedeschi, T. Arakawa, and J. F. Carpenter, Dehydration-induced conformational transitions in proteins and their inhibition by stabilizers. Biophys. J. 65:661–671 (1993).

265. S. D. Allison, A. Dong, and J. F. Carpenter, Counteracting effects of thiocyanate and sucrose on chymotrypsinogen secondary structure and aggregation during freezing, drying, and rehydration. Biophys. J. 71:2022–2032 (1996).

266. H.-T. Yu and B. H. Jo, Comparison of protein structure in crystals and in solution by laser raman scattering. I. Lysozyme. Arch. Biochem. 156:469–474 (1973).

267. P. L. Poole and J. L. Finney, Sequential hydration of a dry globular protein. Biopolymers 22:255–260 (1983).

268. P. L. Poole and J. L. Finney, Sequential hydration of dry proteins: A direct difference IR investigation of sequence homologs lysozyme and α-lactoalbumin. Biopolymers 23:1647–1666 (1984).

269. S. P. Schwendeman, et al., Stabilization of tetanus and diphtheria toxoids aganist moisture-induced aggregation. Proc Natl. Acad. Sci. U.S.A. 92:11234–11238 (1995).

270. H. R. Costantino, Moisture-induced aggregation of lyophilized insulin. Pharm. Res. 11:21–29 (1994).

271. H. R. Costantino, K. Griebenow, P. Mishra, R. Langer, et al., Fourier-transform infrared spectroscopic investigation of protein stability in the lyophilized form. Biochim. Biophys. Acta 1253:69–74 (1995).

272. H. R. Costantino, L. Shieh, A. M. Klibanov, and R. Langer, Heterogeneity of serum albumin samples with respect to solid-state aggregation via thiol–disulfide interchange—implications for sustained release from polymers. J. Controlled Release 44:255–261 (1997).

273. F. MacRitchie, Chemistry at interfaces. San Diego, CA: Academic Press, 1990.
274. T. L. Donaldson, E. F. Boonstra, and J. M. Hammond. Kinetics of protein denaturation at gas–liquid interfaces. J. Colloid Interface Sci. 74: 443 (1980).
275. H. L. Levine, T. C. Ransohoff, R. T. Kawahata, and W. C. McGregor. The use of surface tension measurement in the design of antibody-based product formulations. J. Parenter. Sci. Technol. 45:160–165 (1991).
276. J. Wang and J. McGuire, Surface tension kinetics of the wild type and four synthetic stability mutants of T4 phage lysozyme at the air–water interface. J. Colloid Interface Sci. 185:317–323 (1997).
277. A. F. Henson, J. R. Mitchell, and P. R. Musselwhite, The surface coagulation of proteins during shaking. J. Colloid Interface Sci. 32:162 (1970).
278. C. W. N. Cumper, The surface chemistry of proteins. Trans. Faraday Soc. 46: 235 (1950).
279. W. J. Wang and M. A. Hanson, Parenteral formulations of proteins and peptides. J. Parente. Sci. Technol. 42:suppl. 2S (1988).
280. L. Feng and J. D. Andrade, Protein adsorption on low temperature isotropic carbon. V. How is it related to its blood compatibility? J. Biomater. Sci. Polym. 7:439–452 (1995).
281. M. V. Sefton and G. M. Antonacci, Adsorption isotherms of insulin onto various materials. Diabetes 33:674–680 (1984).
282. W. D. Lougheed, H. Woulfe-Flannagen, J. R. Clement, and A. M. Albisser, Insulin aggregation in artificial delivery systems. Diabetologia 19:1–9 (1980).
283. D. E. James, A. B. Jenkins, E. W. Kraegen and D. J. Chisholm, Insulin precipitation in artificial infusion devices. Diabetologia 21:554–557 (1981).
284. L. Peterson, J. Caldwell, and J. Hoffman, Insulin adsorbance to polyvinylchloride surfaces with implications for constant-infusion therapy. Diabetes 25:72–74 (1976).
285. S. T. Tzannis, W. J. M. Hrushesky, P. A. Wood, and T. M. Przybycien, Adsorption of a formulated protein on a drug delivery device. J. Colloid Interface Sci. 189:216–228 (1997).
286. Z. J. Twardowski, K. D. Nolph, T. J. McGray, and H. L. Moore, Influence of temperature and time on insulin adsorption to plastic bags. Am. J. Hosp. Pharm. 40:583–586 (1983).
287. Z. J. Twardowski, et al., Insulin binding to plastic bags: A methodologic study. Am. J. Hosp. Pharm. 40:575–579 (1983).
288. D. E. Dong, J. D. Andrade, and D. L. Coleman, Adsorption of low density lipoproteins onto selected biomedical polymers. J. Biomed. Mater. Res. 21: 683–700 (1987).
289. K. Matsuno, R. V. Lewis, and C. R. Middaugh, The interaction of γ-crystallins with model surfaces. Arch. Biochem. Biophys. 291:349–355 (1991).
290. J. D. Andrade and V. Hlady, Plasma protein adsorption: The big twelve, Ann. N.Y. Acad. Sci. 516:158–172 (1987).
291. J. H. Lee, J. Kopecek, and J. D. Andrade, Protein-resistant surfaces prepared

by PEO-containing block copolymer surfactants. J. Biomed. Mater. Res. 23: 351–368 (1989).

292. J. H. Lee, J. Kopeckova, and J. D. Andrade, Surface properties of copolymers of alkyl methacrylates with methoxy(polyethylene oxide) methacrylates and their applications as protein-resistant coatings. Biomaterials 11:455–464 (1990).

293. D. Beyer, et al., Reduced protein adsorption on plastics via direct plasma deposition of triethylene glycol monallyl ether. J. Biomed. Mater. Res. 36:181–189 (1997).

294. B. S. Chang, B. S. Kendrick, and J. F. Carpenter, Surface-induced denaturation of proteins during freezing and its inhibition by surfactants. J. Pharm. Sci. 85: 1325–1330 (1996).

295. G. Stempfer, B. Höll-Neughbauer, E. Kipetzki, and R. Rudolf, A fusion protein designed for noncovalent immobilization: Stability, enzymatic activity, and use in an enzyme reactor. Nat. Biotechnol. 14:481–484 (1996).

296. G. Rialdi and E. Battistel, Decoupling of melting domains in immobilized ribonuclease A. Proteins: Struct. Funct. Gene 19:120–131 (1994).

297. G. T. Hermanson, A. K. Mallia, and P. K. Smith, Immobilization affinity ligand techniques. San Diego, CA: Academic Press, 1992.

298. S. Koppenol and P. S. Stayton, Engineering two-dimensional protein order at surfaces. J. Pharm. Sci. 86:1204–1209 (1997).

299. M. N. Gupta, Thermostabilization of proteins. Biotechnol. Appl. Biochem. 14: 1–11 (1991).

300. P. L. Privalov and S. A. Potekhin, Scanning microcalorimetry in studying temperature-induced changes in proteins. Methods Enzymol. 131:4 (1986).

301. E. Freire, Differential scanning colorimetry. In: B. A. P. Shirley, ed. Protein Stability and Folding, Vol. 40. Totowa, NJ: Humana Press, 1995.

302. V. L. Shnyrov, J. M. Sanchz-Ruiz, B. N. Boiko, G. G. Zhadan, and E. A. Permyakov. Applications of scanning microcalorimetry in biophysics and biochemistry. Thermochim. Acta 302:165–180 (1997).

303. L. Swing and A. D. Robertson, thermodynamics of unfolding for turkey ovomucoid third domain: Thermal and chemical denaturation. Protein Sci. 2:2037–2049 (1993).

304. C. N. Pace, F. Vajdos, L. Fee, G. Grimsley, and T. Gray, How to measure and predict the molar adsorption coefficient of a protein. Protein Sci. 4:2411–2423 (1995).

305. W. J. Becktel and J. A. Schellman, Protein stability curves. Biopolymers 26: 1859–1877 (1987).

306. J. H. Carra and P. L. Privalov, Thermodynamics of denaturation of staphylococcal nuclease mutants: An intermediate state in protein folding. FASEB J. 10: 67–74 (1996).

307. G. T. DeKoster and A. D. Robertson, Thermodynamics of unfolding for Kazal-type serine protease inhibitors: Entropic stabilization of ovomucoid first domain by glycosylation. Biochemistry 36:2323–2331 (1997).

308. F. Franks, Accelerated stability testing of bioproducts: Attractions and pitfalls. Trends Biotechnol. 12:114–117 (1994).

309. M. R. Eftnik and R. Ionescu, Thermodynamics of protein unfolding: Questions pertinent to testing and validity of the two-state model. Biophys. Chem. 64: 175–197 (1997).

310. B. A. Shirley, Protein conformational stability estimated from urea, guanidine hydrochloride, and thermal denaturation curves. In: T. J. Ahern and M. C. Manning eds. Stability of Protein Pharmaceuticals. Part A: Chemical and Physical Pathways of Protein Degradation. New York: Plenum Press, 1992, pp. 167–194.

311. T. H. Nguyen and C. Ward, In: Y. J. Wang and R. Pearlman, eds. Stability and Characterization of Protein and Peptide Drugs. New York: Plenum Press, 1993.

312. W. E. Stites, M. P. Byrne, J. Aviv, M. Kaplan, and P. M. Curtis, Instrumentation of automated determination of protein stability. Anal. Biochem. 277:112–122 (1995).

313. D. O. Alonso, K. A. Dill, and D. Stigter, The three states of globular proteins: Acid denaturation. Biopolymers 31:1631–1649 (1991).

314. T. B. L. Kirkwood, Design and anlysis of accelerated degradation tests for the stability of biological standards. III. Principles of design. J. Biol. Stand. 12: 215–224 (1984).

315. U. S. Food and Drug Administration, Points to consider in the manufacture and testing of monoclonal antibody products for human use. J. Immunother. 20: 214–243 (1997).

316. G. Andreotti, et al., Stability of a thermophilic TIM-barrel enzyme: Indole-3-glycerol phosphate synthase from the thermophilic Archaeon sulfolobus solfataricus. Biochem. J. 323:259–264 (1997).

317. H. Tokumitsu, et al., Degradation of a novel tripeptide, *tert*-butoxycarbonyl-Tyr-Leu-Val-CH₂Cl, with inhibitory effect on human leukocyte elastase in aqueous solution and in biological fluids. Chem. Pharm. Bull. Tokyo 45:1845–1850 (1997).

318. K. Patel and R. T. Borchardt, Chemical pathways of peptide degradation. III. Effect of primary sequence on the pathways of deamidation of asparaginyl residues in hexapeptides. Pharm. Res. 7:787–793 (1990).

319. K. C. Lee, Y. J. Lee, H. M. Song, C. J. Chun, and P. P. DeLuca, Degradation of synthetic salmon calcitonin in aqueous solution. Pharm. Res. 9:1253–1256 (1992).

320. A. S. Kearney, S. C. Mehta, and G. W. Radebaugh, Aqueous stability and solubility of CI-988, a novel "dipeptoid" cholescystokinin-B receptor antagonist, Pharm Res. 9:1092–1095 (1992).

321. M. G. Motto, P. F. Hamburg, D. A. Graden, C. J. Shaw, and M. L. Cotter, Characterization of the degradation products of luteinizing hormone releasing hormone. J. Pharm. Sci. 80:419–423 (1991).

322. S. Yoshioka, Y. Aso, and K. Izutsu, Is stability prediction possible for protein drugs? Denaturation kinetics of β-galactosidase in solution. Pharm. Res. 11: 1721–1725 (1994).

323. M. L. Williams, R. F. Landel, and J. D. Ferry, J. Am. Chem. Soc 77:3701–3707 (1995).

324. M. Peleg, On modeling changes in food and biosolids at and around their glass transition temperatures. Crit. Rev. Food Sci. Nutr. 36:49–67 (1996).

325. M. L. Roy, M. J. Pikal, E. C. Rickard, and A. M. Maloney, The effects of formulation and moisture on the stability of a freeze-dried monoclonal antibody—*Vinca* conjugate: A test of the WLF glass transition theory. Dev. Biol. Stand. 74:323–340 (1992).

326. S. Iida and T. Ooi, Titration of ribonuclease T1. Biochemistry 8:3897–3902 (1969).

3
Analytical Methods and Stability Testing of Biopharmaceuticals

Helmut Hoffmann
Boehringer Ingelheim Pharma KG
Biberach an der Riss, Germany

I. INTRODUCTION

Proteins and monoclonal antibodies derived from recombinant DNA are of increasing economic importance for the pharmaceutical industry. More than 30 protein-based pharmaceuticals are presently on the market, reaching total sales of more than $10 billion in 1996 with increasing annual growth rates. Presently about 500 biopharmaceuticals are under development, 200 of which are under clinical investigation (1). Biopharmaceuticals are developed and applied for a broad spectrum of applications, such as tumor therapy and diagnostics, AIDS and other immunological disorders, infectious diseases, neurology, cardiovascular diseases, hematology, wound healing, ophthalmics, skin disorders, diabetes, and respiratory diseases. Besides therapeutic applications, protein-based biopharmaceuticals are applied as vaccines, as drug carriers, and as diagnostic tools (in vivo and in vitro).

A large variety of protein classes are currently being investigated for use as drugs or devices including (recombinant humanized) monoclonal antibodies, hormones, growth factors, interleukins, immune modulators, blood factors, enzymes, and soluble receptors. Access to these proteins has been made available by hybridoma and recombinant DNA technology, supported by the methods of modern biotechnology (2). These technologies have made it economically feasible to manufacture large amounts of complex proteins,

71

available formerly only from natural sources, with the high quality and purity necessary for a modern pharmaceutical product (3).

The appropriate production system for such protein drugs depends on the nature of the desired protein. Bacteria as host cells for recombinant DNA are suitable for nonglycosylated proteins with low molecular weight and a small number of disulfide bridges. Yeast cells and especially mammalian cell culture systems are more appropriate for the production of more complex glycoproteins with high molecular weight and numerous disulfide bridges (4).

Proteins as drug substances differ from conventional chemical drugs with respect to many properties such as size, shape and conformation, multiplicity of functional groups, amphotericity, physical form of bulk drug, and heterogeneity of structure. Therefore, proteins represent complexity that is an order of magnitude greater than that of traditional drugs. This complexity of protein drugs has an impact on the number of analytical methods that must be applied for protein characterization and for the development of stability-indicating assays. In spite of these differences, the general principles of chemical drug stability testing are also applicable to protein drugs. The requirements and the goals are the same: the drugs must be designed to withstand long periods of transport and storage. The full effectiveness and safety of industrially manufactured drugs must be guaranteed until the end of the declared shelf life. This can be achieved and ensured only by development of a stable formulation, supported by extensive product characterization and an appropriate stability testing program.

This chapter describes the methods used to characterize protein drugs and to support formulation development and stability testing. This presentation of methods also focuses on the criteria an analytical method must meet to be stability indicating and useful in screening formulations and in formal stability programs. Because the process used to produce a biopharmaceutical often has a strong influence on the resulting characteristics and stability of the protein, we begin with a brief discussion of the manufacturing process. The discussion focuses on the importance of producing a consistent product, which is achieved by strict control of the cell culture and purification processes.

II. THE IMPORTANCE OF A WELL-CONTROLLED MANUFACTURING PROCESS

The basis for protein stability testing is a comprehensive product characterization with the emphasis on homogeneity of preparation and demonstration of lot-to-lot consistency. Pharmaceutical products are tested for pharmacokinetic parameters, efficacy, and safety during clinical studies. For proteins with structural heterogeneity due to posttranslational modifications (e.g., glycoforms), the degree and profile of this heterogeneity needs to be characterized to ensure lot-to-lot consistency (5). If a consistent pattern of product heterogeneity can be demonstrated throughout development, resulting in a consistent product quality compared to lots used in preclinical and clinical studies, an evaluation of the activity, efficacy, and safety (including immunogenicity) of the individual forms may not be necessary.

Lot-to-lot consistency of biopharmaceuticals can be ensured by validated and well-controlled biotechnical production processes (6). Adequate design of a process and knowledge of its capabilities are part of the strategy used to develop a manufacturing process that is controlled and reproducible, yielding a drug substance/drug product that meets specifications. The manufacturing of the active ingredient comprises a large number of different steps during cell culture, scale-up, fermentation, and protein recovery (7–9). Because so many different process steps that can influence product quality and because protein products are so complex, the quality of a biopharmaceutical is always linked to the process by which it is produced. Therefore, the manufacturing process must be tightly controlled by defining standard procedures for all operations and setting limits for all critical process parameters.

The ultimate stability of a protein can often be a function of the physical and chemical conditions to which it was exposed during processing. A number of critical components of the cell culture process may influence product quality and stability: media conditions, the producing organism, the type of fermentation technology employed, and the techniques used for cell separation and harvesting to isolate the protein prior to the start of purification. To minimize the effects of the cell culture process on product quality and stability, the cells used for production are grown under defined conditions in a fixed and validated fermentation process. After separation from the producing organisms, the protein is subjected to a series of filtration and chromatography steps to remove other protein contaminants. The aim of the first steps in purification is to transfer the desired protein to a high purity, stable form within a short period of time, to avoid potential degradation due to proteolysis. Stability of product intermediates needs to be demonstrated throughout downstream processing to rule out potential degradation that can be caused by such steps as the ultrafiltration, microfiltration, and concentration steps, and by elution conditions during chromatography. Later stages of the purification process serve to remove trace protein impurities and DNA in addition to removing/inactivating potential viral contaminants. For most biologics, final bulk formulation is the last step in the protein recovery scheme, and at this point the purified protein is exchanged into the desired buffer with all appropriate excipients that have been chosen based on the formulation development program. Drug product manufacturing usually encompasses the filling of the formulated bulk protein into final product containers, with subsequent lyophilization if required (e.g., for reasons of stability, drug delivery etc).

During the development phase of the production process, analytical methods are being developed, optimized, and applied on a routine basis to detect deviations of a given protein structure. In many cases, the results of these methods reflect the heterogeneity of the protein—for example, in the profile of a chromatogram or in the banding pattern of an electrophoretic gel. A reference standard is established, which is representative of the product that has been tested in preclinical and clinical trials. This reference material is used in many of the analytical techniques as a basis for comparison, to demonstrate product consistency during process development and scale-up.

After scale-up to the commercial scale, three to five consistency runs are performed, including process validation, to demonstrate the reliability and reproducibility of the process. The product from these consistency runs is used in clinical phase III studies, and the reference standard is established with material from these lots. An extensive analytical characterization of reference material from these consistency runs is performed, and a consistent product quality must be demonstrated for material from these runs. The final product from the consistency runs is put on real-time stability, and the results are used to define the shelf life of the product. All the data obtained from the consistency runs form the basis for approval of the process and product by regulatory authorities.

The concept of lot-to-lot consistency is based on the goal of obtaining identical material from each production cycle, to ensure that the commercial lots possess the same safety, efficacy, and stability profile as the lots that were used in the clinical trials. This goal can be achieved by the establishment of a controlled and validated manufacturing process and by an extensive characterization of the protein drug during quality control and lot release testing. In addition, extensive stability testing is performed to establish the expiration date and guarantee full effectiveness and safety throughout the declared shelf life of the biopharmaceutical product.

III. ANALYTICAL METHODS FOR PROTEIN CHARACTERIZATION OF THE DRUG SUBSTANCE

The characterization of a protein drug is a complex undertaking, requiring the use of a wide range of methods to establish such properties of the drug substance as structural integrity, consistency, activity, purity, and safety (Fig. 1). The complexity of protein molecules means that there are many potential degradation pathways, each with its individual dependences on such parameters as pH, ionic strength, and temperature. Each protein may represent a unique combination of such pathways and dependences. It is therefore critical that a broad spectrum of methods be used to evaluate the effects of processing and storage to assure optimal maintenance of safety and efficacy of the drug (10).

The characterization of a protein drug substance, formulation experiments, and stability testing of a given biopharmaceutical starts when sufficient quantities of highly purified protein are available. The analytical methods applied should have the capability to detect and quantitate different forms of the active ingredient from each other and from their degradation products. Frequently, the protein preparation can be purified to the point of showing only one component by many analytical techniques. However, large and complex proteins, often exhibit "microheterogeneity," which in the case of glycoproteins is due to a number of very similar molecules that possess the same amino acid sequence but differ in the details of their glycan structures. The larger and more complex a protein molecule, the more difficult it is to separate the major product from minor degradation forms by large-scale purification. It is therefore crucial to optimize the analytical methods for maximum resolution; this requires a thorough understanding of the principles of each method and the influence of the operating parameters on method performance (11). In case

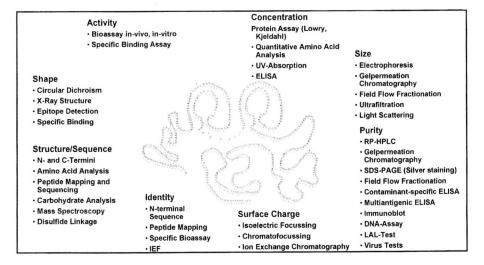

Activity
· Bioassay in-vivo, in-vitro
· Specific Binding Assay

Concentration
Protein Assay (Lowry, Kjeldahl)
· Quantitative Amino Acid Analysis
· UV-Absorption
· ELISA

Shape
· Circular Dichroism
· X-Ray Structure
· Epitope Detection
· Specific Binding

Size
· Electrophoresis
· Gelpermeation Chromatography
· Field Flow Fractionation
· Ultrafiltration
· Light Scattering

Purity
· RP-HPLC
· Gelpermeation Chromatography
· SDS-PAGE (Silver staining)
· Field Flow Fractionation
· Contaminant-specific ELISA
· Multiantigenic ELISA
· Immunoblot
· DNA-Assay
· LAL-Test
· Virus Tests

Structure/Sequence
· N- and C-Termini
· Amino Acid Analysis
· Peptide Mapping and Sequencing
· Carbohydrate Analysis
· Mass Spectroscopy
· Disulfide Linkage

Identity
· N-terminal Sequence
· Peptide Mapping
· Specific Bioassay
· IEF

Surface Charge
· Isoelectric Focussing
· Chromatofocussing
· Ion Exchange Chromatography

Fig. 1 Analytical methods for characterization of proteins.

of product heterogeneity, the different product forms must be characterized, and batch-to-batch consistency of this heterogeneity needs to be demonstrated for production batches (12).

We now turn to a detailed description of the wide range of analytical methods typically used to characterize protein drug substances. The application of analytical methods to formulation development and stability testing is discussed later in the chapter.

A. Primary Structure and Sequence

The amino acid sequence of a given recombinant protein can be derived from the nucleic acid sequence of the gene in the expression vector. The verification of the amino acid sequence on the protein level is most often achieved by peptide mapping and characterization (sequencing) of the isolated peptides by mass spectroscopic (MS) analysis. Peptide mapping is a very powerful and widely applicable tool for protein characterization. The high specificity of certain proteolytic enzymes in cleaving polypeptide chains only at certain residues results in a very characteristic set of peptides (13). Trypsin, for example, cleaves proteins only at lysine and arginine residues, resulting in peptides that end either with lysine or arginine (except the C-terminal peptide), and the same pattern will be obtained for each digest of a given protein. These peptides are then resolved by reversed phase high performance liquid chromatography (RP-HPLC). The HPLC pattern, therefore, results in a "fingerprint" that is characteristic for a given protein. The individual peptides of a protein can be isolated from the HPLC profile and further characterized by mass spectroscopic analysis. This method has the capability to detect changes in the primary structure at a 5–10% level for large proteins such as tissue plasminogen activator (t-PA) (14) and below 5% for smaller proteins such as bovine growth hormone (bGH) (15). Compared to earlier methods for peptide identification and sequencing, such as amino acid analysis and sequencing (see below), peptide identification by mass spectrometry techniques is much faster and more generally applicable. General strategies for analyzing the primary structure of proteins derived from recombinant DNA have been developed on the basis of fast atom bombardment (FAB) MS, allowing precise molecular weight determination of injected peptides (16,17). The utility of the MS method was further increased by the development of an LC/MS interface, such as microbore HPLC with continuous flow FAB, enabling the online sequencing of individual peptides from complex mixtures (18). Such tools permit the confirmation of the primary structure of recombinant proteins in a relatively short period of time (several weeks).

Peptide sequencing is often used to identify the site of a chemical degradation or enzymatic cleavage in degradation products generated during processing or storage of the biopharmaceutical product. Direct N-terminal sequencing has long used a chemical procedure known as Edman degradation, which derivatizes the amino-terminal amino acid in such a way as to release and identify the amino acid and expose the amino terminal to the next cycle of degradation (19). This method has been applied to the intact protein for a limited number of residues, primarily to confirm the integrity of the N-terminus of the protein. In a similar way carboxy-terminal sequencing has been applied for confirmation of the identity/integrity of the carboxy terminus for final products. Today, the identity/integrity of the N-terminal and C-terminal ends of proteins is most often confirmed by identifying and tracking the corresponding peptides in the peptide maps, which are routinely applied for release testing of the products (20). Mass spectrometry is now playing a major role in peptide sequencing, and since there is no problem with N-terminally blocked sequences, MS has the advantage over Edman degradation (21).

B. Secondary and Tertiary Structure

The secondary and tertiary structure of proteins must be considered to be equally important in the maintenance of the overall native status of the protein, and changes in the conformation of proteins may lead to aggregation, decrease in biological activity, and in some cases immunogenicity. Therefore, the design of a formulation for optimal stability of proteins must consider all aspects of the structure of a biopharmaceutical. Methods for evaluating secondary structure are primarily physical, spectroscopic methods and their application to bovine growth hormone has been published (22).

The secondary structure can be evaluated by a technique known as circular dichroism (CD) (23). Amino acids (except glycine) are asymmetric owing to the presence of the chiral carbon. The optical properties of polypeptides are due to the asymmetric centers of its constituent amino acids. Polypeptides, thus, interact differently with right- and left-circularly polarized light. CD is a technique that measures the unequal absorption of left- and right-circularly polarized light. In the far-UV region (<250 nm), CD spectroscopy can be used for the prediction of secondary structures in a protein that are expressed as percentages of α-helix, β-pleated sheet, and random structures (24,25). Because of the sensitivity of signals to detect changes in the environment of the aromatic amino acids (Trp, Tyr, Phe) changes in the tertiary structure of proteins are observed in the near-UV CD region (240–320 nm) (26). Disulfide bonds are also chromophores, which can give rise to CD bands in this near-

UV region, while free thiol (SH—) groups do not (23,27). The near-UV CD spectrum is commonly used as a "fingerprint" of the tertiary structure to provide confirmation that the correct structure is present (28) or that refolding efforts were successful when the product was isolated from an insoluble starting material (29).

Fourier transform infrared spectroscopy (FT-IR) has also provided an estimate of secondary structure composition (30–32). This method uses special deconvolution methods to separate and integrate overlapping amide I infrared absorption bands associated with α-helix, β-pleated sheet, and random structures. In this method, the spectrum is related to the subtle effects of regular secondary structure on the energetics (i.e., vibrational frequency) of amide groups in the peptide linkage. FT-IR also has the advantage of being able to evaluate the aspects of protein structure in the solid state (33). Both this method (34) and calorimetry (35) have been used to study protein conformation at various stages of the freeze-drying process. This information has been used to optimize formulations and processes by maintaining the native structure of the protein.

Many proteins exhibit fluorescence in the 300–400 nm range when excited at 250–300 nm, as a result of the presence of aromatic amino acids (Trp, Tyr, Phe). Fluorescence spectroscopy can yield information regarding the microenvironments of these aromatic amino acids (35). Thus a buried tryptophan is usually in a hydrophobic environment and will fluoresce with a wavelength maximum in the 325–330 nm range, while an exposed residue (or free amino acid) fluoresces at around 350–355 nm (36). Unfolding of a given protein may result in a shift of the fluorescence spectrum due to conformational changes that induce modified emission patterns of the aromatic amino acids in their modified environment.

C. Protein Glycosylation

Glycosylation of proteins is one of the most common and important posttranslational modifications found in eukaryotic secretory proteins. The type and extent of N-glycosylation contributes to the physicochemical and recognition properties of glycoproteins (37,38). The biological activity of glycoprotein hormones frequently depends on the attached N-linked oligosaccharides (39,40). In the recent years, the role of N-glycosylation in human proteins expressed in Chinese hamster ovary (CHO) cells (the most commonly used eukaryotic cells for expression of human recombinant proteins) has been studied extensively (41,42). In many glycoproteins the oligosaccharides contribute to solubility (43) and influence the in vivo circulatory lifetime of the product

(44,45). Other possible functions of the oligosaccharides, such as facilitating secretion, affecting biological activity, or increasing stability, must be investigated for each individual protein. Although N-glycosylation is mainly governed by the type of host cell and the primary structure of the expressed glycoprotein, environmental factors also influence glycosylation (46–50). Alterations in oligosaccharide structures occur either by affecting intracellular synthesis or by changing glycosidase activity after secretion (51,52). Because changes in the glycosylation pattern may have important consequences in glycoprotein pharmaceuticals, carbohydrate analysis is of great importance in product characterization and batch analysis to ensure consistency in product quality.

The microheterogeneity of glycoproteins frequently complicates the interpretation of results from analytical methods for product characterization. Complex carbohydrate structures often contain a varying number of sialic acid residues, which result in a heterogeneity of charge distribution that is manifested by several bands on an isoelectric focusing (IEF) gel or multiple peaks in ion exchange chromatography. The charge heterogeneity of glycopeptides that have been isolated by RP-HPLC after enzymatic cleavage of the protein (peptide mapping) can be separated into individual peaks by capillary electrophoresis.

Substantial progress in the analysis of the oligosaccharide structures of recombinant proteins was made with the introduction by Townsend and Hardy (53,54) of high pH anion exchange chromatography with pulsed amperometric detection (HPAEC-PAD). This technology is superior to refractive index (RI) detection or radioactive labeling of carbohydrates and has emerged as a valuable tool in many laboratories for the characterization of carbohydrate structures of biopharmaceuticals. Different oligosaccharide structures removed from the protein with enzymes such as N-glycanase are separated by HPAEC-PAD after desialylation (failure to desialylate can lead to impaired resolution). The resulting characteristic glycosylation patterns for individual glycoproteins (fingerprints) consist of complex-type bi-, tri-, and tetra-antennary, partially fucosylated, hybrid-type, and high-mannose-type oligosaccharides. A new HPLC-based separation of 2-aminobenzamide (2-AB)-labeled, desialylated oligosaccharides has resulted in better performance and higher sensitivity (5). This HPLC mapping procedure, which can detect even small changes in protein glycosylation patterns, separates different structures that coelute using HPAEC-PAD and is particularly suitable for routine testing of glycosylation consistency in glycoprotein pharmaceuticals.

The glycan structures on a protein are generally quite stable and usually do not change under conditions used for formulation and storage of protein pharmaceuticals. Therefore, glycosylation analysis usually does not belong to a standard set of stability-indicating methods. To analyze for underlying degradation processes, however, it may occasionally be beneficial or necessary to remove the carbohydrates and the associated analytical heterogeneity.

D. Protein Concentration

One of the most fundamental measurements made on a protein is the determination of concentration. Peptides and proteins that do not absorb visible light can react with reagents to form colored compounds. The widely used reagent ninhydrin reacts with amino groups of amino acids and peptides to produce an intensely colored product that has a maximum absorbance at 570 nm. A relatively more specific, but less sensitive, method of quantitatively determining proteins is the biuret reaction of copper in a basic solution in the Lowry assay. The resulting blue product is quantitated in the visible region at 540–560 nm, and the assay is linear at microgram protein levels (55). The Bradford colorimetric assay depends on the binding to the product in an acid environment of the dye Coomassie brilliant blue (56). The reaction can take place in just 2 minutes with a good color stability for 1 hour, to allow measurement at 595 nm. These colorimetric methods can provide relative protein concentrations, when compared with defined standards such as BSA that can be compared between different labs. The binding of the reagents may be different for individual proteins, however, and it also can be influenced by buffer conditions.

Ultraviolet absorption spectroscopy is considered to be the most convenient and accurate measure of the protein concentration (57,58). The absorption spectrum of a protein in the UV wavelength range is a net result of absorption of light by the carbonyl group of the peptide bond (190–210 nm), the aromatic amino acids (250–320 nm), and the disulfide bonds (250–300 nm). Protein concentration can be determined from the absorption spectrum for a purified protein on the basis of its specific absorption coefficient (59). This absorption coefficient is derived from a known protein concentration, usually measured on a 1 mg/mL solution at the wavelength maximum near 280 nm. An accurate independent measurement of the protein concentration (for the sample whose spectrum is recorded) is a critical part of the determination of the extinction coefficient and is often performed by quantitative amino acid analysis, or by nitrogen assay (Kjeldahl) or dry weight measurements. Since, however, the amino acid sequence is known for most protein pharmaceuticals, the theoretical extinction coefficient may also be calculated from the content of the aromatic acid components (60). The effects of the protein structure on the UV spectrum are not considered by this theoretical calculation. The magnitude of these effects can be measured by recording the spectrum when the protein is digested with enzymes until the spectrum stabilizes to that of free aromatic amino acids in the resulting peptides, and their concentration can be determined from their known extinction coefficients (61). This spectroscopic method avoids the errors associated with dry weight analysis and amino acid analysis and is very simple with computer-assisted spectrophotometers.

E. Surface Charge

The acidic and basic groups of proteins make them polyelectrolytes, and they can therefore be separated in an applied electric field. The use of this phenomenon, known as electrophoresis, is one of the most common methods of separating mixtures of proteins on an analytical scale. Many different forms of electrophoresis are applied, and details on the theoretical principles behind these electromigration techniques have been published (62,63).

In native electrophoresis, performed in the absence of denaturing agents, the major factors controlling the electrophoretic mobility of a macromolecule are its net charge (at the pH of separation) and its Stokes radius, a hydrodynamic parameter determined primarily by the size and, to a lesser extent, the shape of the molecule. The basis for separation is therefore determined mainly by differences in mass-to-charge ratio. The use of an anticonvective matrix, such as a gel, can enhance the resolving power of this method. If the pores of the gel are comparable to the dimensions of the protein, they will present resistance to the movement of the molecules in a size-dependent fashion. Since native gel electrophoresis provides information on both size and charge, independent data are required to resolve these two factors.

The charge of a native protein is dependent on the pH of the solution. At the isoelectric point (pI), the protein's net charge will be zero and the mobility in the electric field will be zero. Thus, if the electric field is also a pH gradient, the protein will migrate to the point at which the pH is the same as the pI and the migration will stop. This phenomenon is known as isoelectric focusing (64). Stable pH gradients are established by means of carrier ampholytes with appropriate pI and buffering capacity, which are included in large-pored agarose or acrylamide gels (65,66). Such IEF systems can separate protein species that differ in pI by as little as 0.02 pH unit (67). This method has been applied to separate different glycoforms of proteins that differ in the degree of sialylation of the complex carbohydrates (68,69). The method is also widely used as a stability-indicating method to assess deamidation during stability studies by quantitative densitometry (70).

Ion exchange chromatography is a powerful separation technique at the preparative scale as well as an analytical tool for assessment of charge heterogeneities in protein preparations (71). A protein with a net positive charge tends to bind to a matrix with a net negative charge by ionic interactions, and vice versa, provided the ionic strength of the buffer is sufficiently low. The passage of a salt gradient over a column to which a mixture of proteins has bound will cause the elution of each protein at its own critical salt concentration, where the protein binds less tightly than the salts. Thus the proteins, or differently charged variants of a protein, are separated from each other, and the components of the mixture can be individually quantitated. The charge distribution on the surface of a given protein is dependent on the solution pH, which dictates the binding and elution conditions of the ion exchange chromatography method. If two forms of the same protein, which differ by one charged amino acid, are analyzed at a pH where that group is uncharged, this difference may not be detected by ion exchange chromatography. Therefore, a single homogeneous peak on ion exchange is not a guarantee that the sample is homogeneous.

Chromatofocusing, as the name implies, combines aspects of isoelectric focusing and ion exchange chromatography. It is performed under low salt conditions and can use the same ampholytes used in isoelectric focusing (72,73). The sample is loaded onto an ion exchange column at a pH where it binds. The column is then eluted with an ampholyte or buffer mixture selected to generate a pH gradient that gradually flows down the column. When the pI of a bound protein is reached, that protein is released from the resin. The method has high resolving power and the proteins are eluted in highly focused peaks from the column. This method has one significant disadvantage because most proteins exhibit minimum solubility at low ionic strength around their pI, which means that neutral surfactants or urea may be required to keep the protein in solution.

F. Protein Size

The most common form of protein electrophoresis is polyacrylamide gel electrophoresis with the denaturing agent sodium dodecyl sulfate (SDS-PAGE). Reduced proteins tend to bind a relatively constant amount of SDS on weight basis: approximately 1.4 of SDS per gram of protein (74). The SDS molecule carries a negative charge, and complexes of proteins with SDS have very similar mass-to-charge ratios and therefore free electrophoretic mobility (75). When a mixture of SDS-saturated proteins is electrophoresed in a gel matrix with the correct pore size, the major factor determining their migration rate is their effective size. SDS-PAGE is commonly used for assessing purity and as a tool for determining the apparent molecular weight of a protein (76,77). The most reliable estimate of molecular weight is obtained from analyses in which the disulfide bonds have been reduced, and the polypeptide chains are truely random (75). A widely used system is the discontinuous buffer system with SDS published by Laemmli (78). This system may be used for proteins ranging in molecular weight from 10,000 to 300,000 Da by varying the concentrations of the acrylamide gel and the cross-linker, bisacrylamide, to vary the pore size of the gel (75,79).

The resolving power of SDS-PAGE can be further increased by using gradients of acrylamide to vary the pore sizes in the gel. Determination of the apparent molecular weight of a protein can be influenced by sample treatment (e.g., heating of samples and use of reducing agents, such as mercaptoethanol or dithiothreitol or carboxymethylation with iodoacetic acid), which then makes the results difficult to interpret. It is also common to find that proteins with significant carbohydrate content running as diffuse bands, probably because of a combination of heterogeneity of molecular weight, charge (as a result of variable extents of sialylation), and SDS binding (80). Protein quantitation by SDS-PAGE can be complicated by the variable dye-binding properties of individual proteins (81,82) and the denaturing effects of SDS.

The apparent molecular weight of a protein can also be determined by gel filtration or size exclusion chromatography on a column that has been calibrated with molecular weight standards (83). This method is of relatively low resolution primarily because it is based on the hydrodynamic properties of the protein and consequently gives accurate estimates only for spherical proteins. Since separation by gel filtration is based on apparent molecular weight, this method is frequently used to assess the aggregation state of the protein and to quantitate dimers, oligomers, and aggregates in a protein preparation (84,85). While yielding lower resolution than SDS-PAGE, this method does allow the quantitation of concentration.

The incorporation of specialized ionization methods has extended the applications of mass spectroscopy to protein characterization and accurate determination of molecular weight: ESI-MS (electrospray ionization mass spectrometry) and MALDI-MS (matrix-assisted laser desorption ionization MS). ESI-MS has extremely high precision because it generates multiple-charged species of even very large proteins (reviewed in Ref. 86). This precision arises from the generation by each peak in the family of peaks generated by the differently charged species of an estimate of the molecular weight (MW), and these estimates can be combined to increase the overall precision of the MW estimate. This approach was used to characterize recombinant γ-interferon and its C-terminal degradation products simultaneously, yielding an estimate of the MW of 16908.4 ± 1.2 while the theoretical mass is 16907.3 (87). MALDI-MS was pioneered by Hillenkamp and Karas (88,89), who showed that if a high concentration of a chromophore is added to the sample, a high intensity laser pulse will be absorbed by the matrix and the energy absorbed will volatize a portion of the matrix, carrying the protein sample with it into the vapor phase essentially intact. The resulting ions are then analyzed in a time-of-flight MS (TOF-MS). The "gentle" nature of the ionization may be responsible for the ability of the method to provide information on quaternary structure (89). A major extension of the TOF-MS method was developed by Beavis and Chait (90), who showed that the method is relatively insensitive to large amounts of buffer salts and iorganic contaminants. This type of methodology may have a wide utility for several reasons: it requires only picomole amounts of sample; it is very fast (<15 min from start to finish) and does not fragment the molecules; the sample can be a crude mixture of proteins; and the result is in principle as easy to interpret as (and indeed resembles) a densitometric scan of an SDS-PAGE gel, with a mass range well above 100 kDa.

G. Bioassay, Potency Assays

All the methods described thus far are universally applicable to different proteins, although the selection and the focus of the analytical methods applied will be influenced by the specific properties of the protein. Potency assays, however, are required to mimic the specific biological activity of the biopharmaceutical, hence are usually protein specific. For many proteins, in vivo bioassays in animals have been developed, to measure the "true" biological activity including pharmacodynamics, of the product. For example, the bioassay for human growth hormone (hGH) measures the daily weight gain in hypophysectomized (hypoxed) rats given daily injections of hGH (91). The rats respond to exogenous growth hormone (even from different species). Usually 10 rats are used per group, and two doses are compared for the sample, with inclusion of a reference standard and a blank as controls. The dosing and the recording phase of the study takes 10 days. As can be seen from the design of this bioassay, the analysis of a single sample requires a tremendous effort. Therefore, such a bioassay is not suitable for analyzing large numbers of samples derived from formulation screening or from stability studies. In addition, such in vivo bioassays have the drawback of animal use and additionally suffer from poor reproducibility. In many cases, cell-culture-based in vitro bioassays that mimic the biological activity of the protein have replaced in vivo bioassays. The murine thymocyte proliferation assay that is widely used for routine analysis of human interleukin 1 (hIL-1) activity is an example of such a surrogate assay (92). The proliferation by hIL-1 is mediated via IL-2 release of hIL-1 stimulated T cells, since antibodies specific for hIL-2 can block this response (93). The assay, which is performed in microtiter plates, is sensitive to 10–50 pg/ml Il-1. Proliferation is assessed by [^{3}H]thymidine incorporation after 72 hours of culture. A major problem with this assay is its lack of specificity: it can be stimulated by hIL-2 and compounds used to induce hIL-1 production (e.g., lipopolysaccharide and phorbol myristate acetate). Similar proliferation assays were developed for hIL-1 on the basis of different cell lines. Three different amino-terminal variants of recombinant hIL-1β (rhIL-1β) were demonstrated to differ in their activity by 3- to 10-fold in these cell-based bioassays (94).

The biological activity of interferons is measured by their dose-dependent inhibition of the cytopathic effect (CPE) due to virus infection of cell cultures in microtiter plates (95). The activity is measured in comparison to international standards and is expressed in international units (IU) (96). Although cell-based in vitro assays have a much higher sample throughput and reproducibility than in vivo assays in animals, the assay variability of the former is still comparatively high, with coefficients of variation in the range of 20–50%.

For some proteins, especially for enzymes, biomimetic in vitro test systems with good reproducibility have been developed. An example is recombinant tissue plasminogen activator (rt-PA), a serine protease that cleaves plasminogen to plasmin and thereby initiates the lysis of fibrin clots and blood coagulates. This biopharmaceutical is used for several indications, such as myocardial infarction, stroke, lung embolism, and deep venous thrombosis. The enzyme activity of rt-PA can be measured in a chromogenic assay using a synthetic substrate, a tripeptide linked to *p*-nitroanilide (S-2288). The rate of cleavage is monitored following the formation of *p*-nitroaniline spectrophotometrically at 405 nm. The one- and two-chain forms of rt-PA have different affinities for the substrate and therefore differ in their specific enzymatic activity in this assay (64). The concentrations of one- and two-chain rt-PA can be determined based on the difference in amidolytic activity between the two forms toward S-2288 substrate. For this purpose the assay is performed with the mixture of one- and two-chain rt-PA and also after all the rt-PA in the mixture has been transformed to the two-chain form (with higher specific enzymatic activity) by adding a trace amount of plasmin to the mixture.

A more specific and relevant enzyme assay is an indirect chromogenic assay for rt-PA in which a synthetic substrate (S-2251) specific for plasmin is used to measure the plasmin generated upon incubation of rt-PA and plasminogen. The amount of plasmin generated in this test correlates with the plasminogen-activating potency of t-PA (97). The most biologically relevant in vitro potency assay for rt-PA is the in vitro clot lysis assay, which is based on measuring the time taken for a fixed amount of rt-PA to dissolve a fibrin clot. Typically, a fibrin clot is produced by combining fibrinogen and thrombin in the presence of plasminogen. Then rt-PA is introduced to initiate the lytic reaction. If rt-PA is present at limiting amounts, the time for clot lysis is directly proportional to its concentration. The accuracy of the assay is dependent on the ability of the analyst to reproducibly measure the reaction end point. Several methods have been used, which include releasing entrapped air bubbles from the clot and dropping glass beads through the fibrin clot (98,99). In an automated version of the in vitro clot lysis assay, end point detection is based on turbidometric measurement performed by means of a commercially available microcentrifugal analyzer (100). Lysis of the clot is followed by measuring the decrease in absorbance at 340 nm. This automated version of the in vitro clot lysis assay is reliable and reproducible, with an accuracy of 99.5% and a precision of 5% (in the concentration range of 40–1200 ng/mL rt-PA) and can be performed at high throughput with minimal sample handling. The in vitro clot lysis assay has also been adapted to a microtiter plate format (101).

Another fibrinolytic assay system measures the lysis of clots prepared from human blood, then incubated in plasma in the presence of small amounts of added rt-PA (102). In this system, the fibrinolytic activity of rt-PA is determined by plotting the loss in blood clot weight against the concentration of rt-PA in the plasma sample. Since this so-called hanging-clot assay is very time-consuming and not suited to automation, its application is limited to very specific investigations; it is not useful for routine quality control purposes.

General requirements for a stability test are high capacity and throughput of samples, as well as high accuracy and precision. Both criteria are fulfilled by the automated in vitro clot lysis assay for rt-PA. This assay is one of the stability-indicating assays for rt-PA and has been used to assign the expiration date (shelf life) of drug product lots.

For many protein drugs, no enzymatic activity can be measured. This is usually the case for monoclonal antibodies that bind to a specific antigen, for soluble receptors or receptor ligands, and for virus subunit proteins. In these cases, a choice for an in vitro potency assay can be a competitive specific binding assay. This competition assay can be designed as a sandwich assay, the soluble antigen fixed to a microtiter plate. The displacement of a conjugated form of the monoclonal antibody is measured for the nonconjugated test sample of the same monoclonal antibody and for a defined reference material of the antibody on the same microtiter plate. From the displacement calibration curve for reference material, a relative binding potency for the sample can be determined. This assay format can be performed with high accuracy, precision, and capacity in sample throughput. However, a specific binding assay does not necessarily reflect the ''true potency'' of a protein drug. Therefore, care must be taken to correlate the data from in vitro binding assays with more relevant biological test systems. In addition, the stability-indicating properties of such an in vitro assay need to be established.

H. Product Purity

The determination of absolute as well as relative purity presents considerable analytical challenges, and the results are highly method dependent. Historically, the relative purity of a biological product has been expressed in terms of specific activity (units of biological activity per milligram of product) which is also highly method dependent. Consequently, both drug substance and drug product are assessed for purity by a combination of methods.

There is an inherent degree of structural heterogeneity in proteins, a result of the biosynthetic processes used by living organisms producing the protein. Therefore, the desired product can be a mixture of posttranslationally modified forms (e.g., glycoforms, as described above). These forms may be active, and their presence may have no deleterious effect on the safety and efficacy of the product. When variants of the desired product are formed during the manufacturing process and have properties comparable to the desired product, they are considered to be product-related substances, not impurities.

Biopharmaceuticals, composed of the desired product and multiple product-related substances, need to be tested for impurities, which may either be process-related or product-related.

Process-related impurities encompass those that are derived from the manufacturing process, classified in three major categories: cell-substrate-derived, culture-derived, and downstream-derived. Impurities derived from cell substrates include host cell proteins, nucleic acid (host cell generic, vector, total DNA), lipids, polysaccharides, and viruses. For host cell proteins, a sensitive immunoassay capable of detecting a wide range of protein impurities is generally utilized. The polyclonal antibodies utilized in the test are generated from a crude preparation of a mock production organism (i.e., a production cell minus the product coding gene). The level of DNA from host cells can be detected by direct analysis of the product using DNA hybridization techniques or a Threshold total DNA assay (103). DNA spiking experiments may be performed at laboratory scale in order to validate efficient removal of DNA during the downstream processing. Culture-derived impurities include media components such as antibiotics, serum and media-derived proteins. Downstream-derived impurities include enzymes, chemical/biochemical processing reagents, inorganic salts, solvents, carrier/ligands (e.g., monoclonal antibodies, protein A), and other leachables.

Product-related impurities, such as degradation products, are molecular variants arising from processing or during storage, which do not have properties comparable to those of the desired product with respect to activity, efficacy, and safety. Product-related impurities may arise via several degradation pathways, described below. For the purpose of stability testing, as well formulation development, the tests for purity mainly focus on methods for determination of degradation products. Therefore, the degree of purity, as well as the amount of individual and total degradation products in the biopharmaceutical product lots entering a stability study, need to be described and documented. The intrinsic purity of a given protein preparation, especially with respect to trace amounts of host-cell-derived proteases in liquid formulations, may have a significant impact on the stability profile of a given biopharmaceutical and on its lot-to-lot variability.

I. Safety Testing

Most biopharmaceutical drugs are delivered parenterally, which requires final products to be tested for mycoplasma, sterility, and pyrogenicity. Pyrogenicity testing may be replaced by Limulus amoebocyte lysate (LAL) testing for endotoxin according to the U.S. Pharmacopeia or European Pharmacopoeia. The General Safety test is another safety test specific to biologics and is intended to detect any unexpected or unwanted biological reactivity with a product. The assay encompasses inoculation of guinea pigs and mice with the final formulated product. The final product lot will pass the test if no unforeseen reactions occur and no weight loss takes place during the test (104).

In addition to final product testing, good manufacturing practices (GMP) are applied during processing to avoid any contamination with adventitiously introduced materials, not intended to be part of the manufacturing process, such as biochemical/chemical materials and/or microbial species (105). Special requirements are applied to products derived from mammalian cell cultures to avoid any contamination with viruses. These include a combination of testing and validation of the downstream purification process for removal and inactivation of adventitious viruses and viruses intrinsic to the cell line

(106,107). Specific guidelines address these virus safety aspects for cell-culture-derived products (108,109).

IV. ESTABLISHING STABILITY-INDICATING ANALYTICAL METHODS FOR FORMULATION DEVELOPMENT AND STABILITY TESTING OF BIOPHARMACEUTICAL PRODUCTS

An extensive physicochemical characterization of the protein, as described above, forms the basis for the development of a stable formulation (for another review, see Ref. 110). The physicochemical properties of the protein, its behavior in different solutions, as well as the purpose and application of its in vivo use, will guide the choice of formulation. The strategy for development of protein formulations is discussed in more detail in Chap. 4. Analytical results from product characterization and first experiences concerning product stability obtained during process development provide a good basis for the selection of stability-indicating test methods. In many cases these tests are product specific assays such as those measuring potency or activity, or cover specific features unique to an individual protein. In addition, some general tests are performed and some general requirements for the scope of stability-indicating test methods for biopharmaceuticals can be defined.

Stability-indicating test methods should detect the most common degradation forms of biopharmaceuticals: inactive or denatured protein, soluble and insoluble aggregates, proteolytically truncated forms, and chemical modifications, such as hydrolysis, deamidation, oxidation, disulfide exchange, β-elimination, and racemization. A more detailed discussion of degradation pathways of proteins is given in Chap. 2 (for a review, see Ref. 111). A summary of the most common degradation pathways and examples of methods used to detect degradation products is given in Table 1.

We now discuss these common degradation pathways, emphasizing the analytical techniques useful for monitoring changes in proteins during stability testing. These methods will be intergral to the design and optimization of stable formulations.

Protein denaturation refers to a disruption of the higher order structure, such as secondary and tertiary structure of a protein. Denaturation, which may be reversible or irreversible, can be caused by thermal stress, extremes of pH, and exposure to interfaces or denaturing chemicals. Denaturation typically

Table 1 Common Degradation Routes[a] and Methods[b] Applied to Detect Degradation Products

Degradation rule	Region affected/Results	Major factors	Method
Aggregation	Whole protein; reversible or irreversible self-association	Shear, surface area, surfactants, pH, T, buffers, ionic strength	Size exclusion chromatography Light scattering Analytical ultracentrifugation
Deamidation	Asn or Gln; acidic product, isoform, or hydrolysis	pH, T, buffers, ionic strength	Isoelectric focusing Ion exchange chromatography Native electrophoresis Reversed-phase HPLC
Cleavage	Asp-X; fragments (proteolysis also possible from trace proteases)	pH, T, buffers	N- and C-terminal sequencing Size exclusion chromatography Reversed-phase HPLC (peptide map) SDS–polyacrylamide gel electrophoresis Isoelectric focusing

Oxidation	Met, Cys, Hid, Trp, Tyr; oxidized forms	Oxygen (ions, radicals, peroxide), light, pH, T, buffers, metals, (surfactants), free radical scavengers	Reversed-phase HPLC (peptide map) Hydrophobic interaction chromatography Amino acid analysis
Thiol disulfide exchange	Cys; mixed disulfides; intermolecular or intramolecular	pH, T, buffers, metals, thiol scavengers	Reversed-phase HPLC (± reduction) Reversed-phase HPLC (peptide map) SDS–polyacrylamide gel electrophoresis
Altered secondary structure	Whole protein	Shear, surface area, surfactants, pH, T, buffers, ionic strength	Far-UV circular dichroism Infrared spectroscopy
Altered tertiary structure	Whole protein	Shear, surface area, surfactants, pH, T, buffers, ionic strength	Near-UV circular dichroism UV absorption spectroscopy Fluorescence spectroscopy

[a] This table lists degradation pathways commonly observed for proteins and peptides. The listing is not comprehensive, however, and many of these degradation routes may occur independently or in combination with one another.
[b] Methods that are frequently used for analysis of alteration.

involves unfolding of the protein. In some cases the unfolded protein can be transformed back to its native state by use of denaturants such as guanidine hydrochloride or urea, followed by dialysis (112). If the protein cannot easily recover its native state by refolding, denaturation is considered to be irreversible. In many cases this leads to aggregation and precipitation phenomena.

Changes in the secondary and tertiary structure of proteins can be monitored by spectroscopic methods, such as UV-CD and FT-IR, as described earlier in this chapter. In some cases unfolded proteins can be separated from their native forms based on differences in chromatographic behavior (e.g., separation by hydrophobic interaction chromatography (HIC) due to an altered hydrophobicity pattern of the protein). Another useful technique for investigating changes in protein conformation as on exposure to specific environments is differential scanning calorimetry (DSC). As a protein is heated, the transition from the native state to the unfolded state is accompanied by the appearance of an endothermic peak on DSC. The transition temperature, T_m, is analogous to the melting of a crystal and is affected by the environmental conditions (e.g., pH) and the presence of pharmaceutical excipients. Sugars such as glucose and sucrose and polyols such as sorbitol and glycerol have been found to increase the denaturation temperature of proteins (113,114). A pH dependency of the endothermic peaks in DSC has been shown for several proteins (68,115). For rt-PA, the melting temperature in phosphate buffer was found to be about 66°C. In the presence of arginine, which stabilized the protein in solution, the T_m shifted to 71°C (116). Figure 2 contrasts a DSC thermogram for rt-PA with a melting temperature of 69.87°C with a scan for an rt-PA variant having a significantly lower T_m (64.20°C). The thermogram for a humanized monoclonal antibody shown in Fig. 3 exhibits multiple transition temperatures for the heavy and light chains in the antibody.

Aggregate formation is one of the most common forms of protein instability (117). Insoluble and soluble aggregates must to be distinguished. Soluble aggregates can be detected and quantitated by gel filtration chromatography. They do not necessarily lead to opalescence or turbidity of the protein solution. In contrast, insoluble aggregates consist of large protein particles that do not enter gel filtration columns and manifest themselves in the form of haziness or opalescence in a solution that is intended to be clear. Turbidity or opalescence in protein solutions can be caused by small amounts of insoluble aggregates, often less than 1%. All final product lots of biopharmaceuticals that are used as parenterals are inspected for appearance and clarity. A spectrophotometric determination of the opalescence of a protein solution in comparison with reference preparations as defined by the European Pharmacopoeia serves as an objective standard of evaluation for the presence of insoluble aggre-

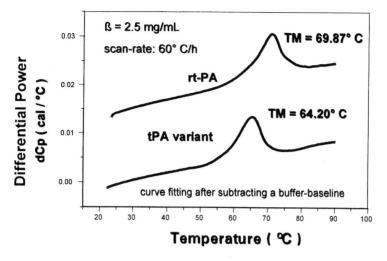

Fig. 2   Differential scanning calorimetry of rt-PA and an rt-PA variant.

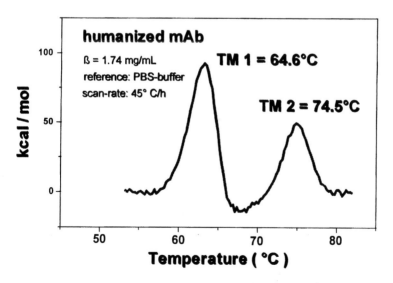

Fig. 3   Differential scanning calorimetry of a humanized monoclonal antibody.

gates. To obtain accurate values when one is quantitating the concentration of a protein solution in the presence of opalescence, protein content measurements made by UV-absorption spectroscopy need to be corrected for increased absorbance, due to light scattering (118).

Insoluble protein aggregates can also be detected and quantitated as particles by light scattering techniques. Particulates in injectables are assessed according to the USP Particulate Matter test and need to meet the acceptance criteria (< 600 particles > 10 µm and < 6000 particles >1 µm). Photon correlation spectroscopy (PCS) is the most useful technique for particle size analysis of submicrometer particulates, having a range of application from a few nanometers (corresponding to the size of proteins in solution) to a few micrometers (119). PCS is a light scattering technique for the measurement of the statistical intensity fluctuations in light scattered from the particles. These fluctuations are due to the random Brownian motion of the particles, which are size dependent. The diffusion coefficients of a protein can be determined from an autocorrelation function of the PCS, allowing particle size assignments. As with other light scattering techniques, the analysis of PCS data becomes more complex when the sample particles are not monodisperse. However, PCS has been used to study the hydrodynamic size of proteins and their dependence on pH (120). The method is suitable for the quantitation of monomers and dimers and has been used to elucidate the refolding process of proteins (121). The measurement of monomers and dimers of proteins is disturbed by small amounts of aggregates, which can be detected at a very high sensitivity by PCS.

In addition to considering aggregation of the protein active ingredient in a formulation, it is important to keep in mind that excipients in a formulation are also capable of aggregation. Proteins formulated at low concentrations may interact with container/closure systems, resulting in losses by absorption (122). In some cases, stabilizers such as human serum albumin (HSA) or surfactants are added to prevent such absorption effects. In one case described for an unstable HSA/dextrose solid formulation at elevated temperatures, however, a broadening and shift of the HSA peak together with a decrease in retention time in RP-HPLC was observed (Fig. 4). This was accompanied by a loss in mobility on SDS-PAGE gels (not shown) and an increase in the molecular weight of HSA as determined by mass spectroscopy analysis (Fig. 5). MS analysis of tryptic peptides of the HSA revealed a modification of lysine residues by a glucosylation of the ε amino group (123). These results strongly suggest an interaction between the HSA and dextrose excipients in the formulation at temperatures above 25°C. The development of HSA-free

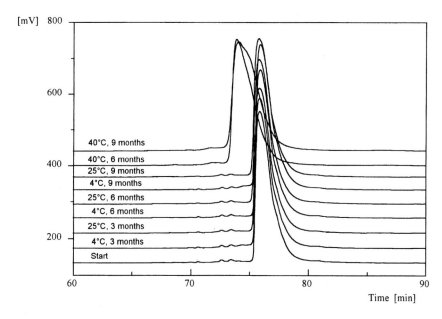

Fig. 4 Stability profile of HSA in a solid dextrose formulation monitored by RP-HPLC.

formulations is recommended even for low dose biopharmaceuticals, however, because HSA is less well defined and less pure compared to recombinant proteins and additionally bears the potential risk of virus contamination due to the source of multidonor blood pools. Such protein excipients may also undergo the same degradation pathways as the therapeutic protein, introducing the need for stability testing here, as well.

Another degradation pathway for proteins is hydrolysis through enzymatic and nonenzymatic routes resulting in cleavage of a peptide bond. Cleavage products can be visualized by SDS-PAGE using sensitive silver staining techniques (124–126). Silver staining techniques are mainly used when sensitivity is a major issue, such as in detection of impurities in pharmaceutical protein preparations. However, the band intensities are not linearly proportional to the amount of protein loaded and the silver binding properties of proteins (127). Thus, quantitation of silver stained gels is problematic. When silver staining is used, the detection limit for proteolytic degradation products of proteins is in the range of 200–1000 parts per million (ppm) for individual

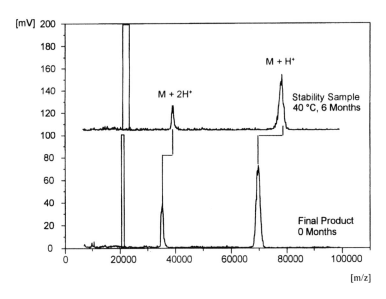

Fig. 5 MALDI TOF mass spectroscopy of HSA in a dextrose formulation, before and after storage for 6 months at 40°C.

bands on a gel. Staining of SDS gels with Coomassie blue is 10 times less sensitive, but more suitable for quantitative measurement of degradation products by densitometric scanning of the gels.

Figures 6 and 7 illustrate results of the quantitative measurement of degradation products during stability testing of a monoclonal antibody by densitometric scanning of SDS gels. In Fig. 6, the densitometric scan profiles from SDS-PAGE gels, main peaks of the heavy and light chains of a reduced antibody are shown, as well as peaks from degradation bands that increase during storage time. Figure 7 is the graphic representation of the combined peak areas for the heavy and light chains and their changes as a function of storage time and temperature.

Proteolytic processing of proteins can also be measured by chromatographic techniques like gel filtration chromatography. Rt-PA has a predominant proteolytic cleavage site at position amino acid 275-276 and is readily cleaved by plasmin into two-chain rt-PA. The ratio of single-chain to two-chain t-PA can be determined by HPLC–gel permeation chromatography performed under reducing conditions. Rt-PA produced under serum-free conditions by mammalian cell culture exists predominantly in the single-chain form.

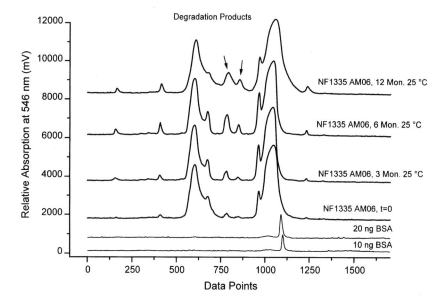

Fig. 6 Stability testing of a monoclonal antibody preparation by SDS-PAGE and densitometric scanning of the gel.

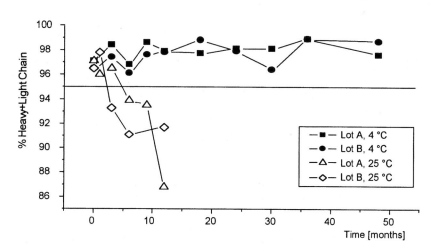

Fig. 7 Densitometric scan evaluation of SDS-PAGE used for stability testing of monoclonal antibody preparations after storage at two different temperatures.

During long-term storage in liquid formulation, this molecule will be cleaved into the two-chain form, depending on storage temperature and pH of the formulation buffer. Gel permeation chromatography is applied for stability testing of rt-PA for both quantitation of aggregates and for quantitation of the truncated two-chain form of the protein (Fig. 8).

In addition to proteolytic cleavage due to the presence of trace amounts of proteases, the peptide bond can undergo nonenzymatic hydrolysis, resulting in protein degradation. The Asp-Pro peptide bond is known to be the most susceptible to hydrolytic breakdown. The preferential hydrolysis of certain Asp-Pro bonds in a protein structure may be due to greater accessibility of the group (128) or to the location of these bonds adjacent to other amino acids that may influence the hydrolytic reaction (129). Degradation products or truncated forms of proteins generated by hydrolysis are analyzed by the same set of methods applied to truncated forms generated by proteolysis.

Oxidation is another of the major chemical degradation pathways of peptides and proteins, both in solution and in lyophilized formulations. Amino acids that may undergo oxidation include methionine, cysteine, histidine, tryp-

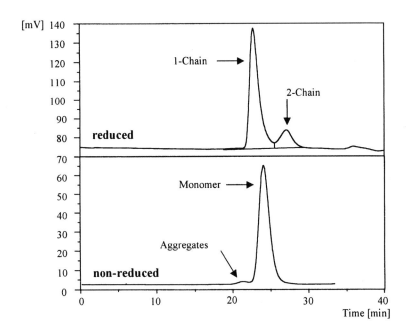

Fig. 8 HPLC-SEC analysis of rt-PA under non-reduced and reduced conditions.

tophan, and tyrosine. Most oxidation reactions commonly encountered in therapeutic proteins under normal storage conditions involve methionine and/or
cysteine residues. Methionine is easily oxidized, even by atmospheric oxygen,
to methionine sulfoxide. The oxidation of methionine is catalyzed by trace
amounts of peroxide or metal ions, which may be present as contaminants
from the manufacturing process or in trace amounts in excipients (44,
112,130). The oxidation of methionine and cysteine is influenced by the three-
dimensional structure of the protein. Residues that are buried in the interior
of the protein are inaccessible to oxidation but can become reactive upon unfolding of the protein (95). Oxidized variants of proteins can be detected and
characterized by liquid chromatographic techniques, such as RP-HPLC or by
peptid mapping, followed by MS analysis. An example of an RP-HPLC separation of product variants including oxidized forms is shown in Fig. 9.

Cysteine amino acid groups are also easily oxidized to yield cysteine
disulfides. During long-term storage, free sulfhydryl groups may be oxidized
to form intrachain or interchain disulfide linkages. Such interchain bonds between multiple protein molecules lead to protein aggregation (131). Under
thermal stress, a protein will often undergo a destruction of the disulfide bonds
by β-elimination from cysteine residues, resulting in free thiols that may contribute to other degradation pathways (112). The generation of free thiols by
β-elimination may, in turn, catalyze disulfide interchange. Disulfide interchange may also result from the presence of unpaired cysteine residues. These

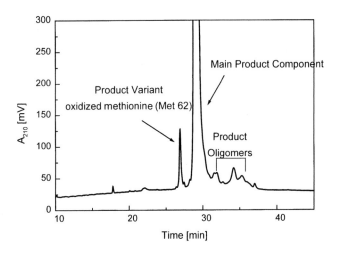

Fig. 9 RP-HPLC separation of product variants.

cysteine residues can react at different sites to form new disulfide bridges, resulting in proteins with incorrect disulfide linkages and nonnative conformation (51,58). Confirmation of the correct linkage of disulfide bridges in a protein is a monumental analytical undertaking and requires cleavage of the protein under reducing and nonreducing conditions and characterization of the isolated peptides by LC-MS to assign the cysteines involved in the disulfide bonds. Improperly folded proteins derived from disulfide interchange may be separated by high resolution chromatographic techniques, such as RP-HPLC, HIC, SEC, or IEC.

Nonenzymatic deamidation is a very common hydrolytic reaction responsible for degradation of peptides and proteins involving the amide group of asparaginyl or glutaminyl (Asn and Gln) residues. Asn and Gln residues are labile at extremes of pH and may be hydrolyzed easily to free carboxylic acids (Asp and Glu, respectively). Deamidation may have significant effects on protein bioactivity, half-life, conformation, aggregation, and/or immunogenicity. These effects must be evaluated on a case-by-case basis, since deamidation does not always affect bioactivity or the half-life of the product. However,

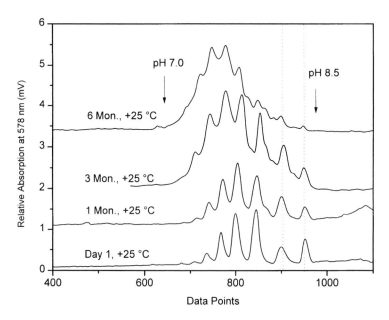

Fig. 10   IEF profiles by densitometric scan analysis of a monoclonal antibody preparation, stored at 25°C in liquid formulation.

deamidation results in a change in the original primary amino acid sequence of the protein, which may then be more susceptible to irreversible aggregation and more rapid clearance (117). Deamidated variants of proteins can be detected and quantitated by IEF, followed by densitometric scanning or by ion exchange chromatography. Figure 10 illustrates the changes observed by IEF for a monoclonal antibody stored for several months at 25°C. With increased storage time, there is a shift in the IEF pattern toward lower pH values, which would be consistent with a deamidation mechanism.

Once methods capable of detecting these common forms of protein degradation have been identified, they must be validated according to current guidelines for specificity, linearity, accuracy, recovery rate, sensitivity, precision, and limit of detection (132). The stability-indicating nature of each individual test method must be established for each individual protein. This process yields a specific stability profile for a given biopharmaceutical. We often determine whether a method is stability indicating by accelerating the degradation of a protein by exposure to elevated temperatures. Figure 11, for example, illustrates the temperature-dependent decrease in binding potency of a monoclonal antibody in a competitive specific binding assay compared to binding of the reference standard.

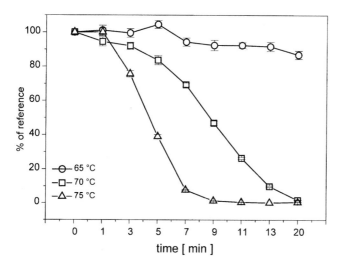

Fig. 11   Accelerated stability testing of a monoclonal antibody preparation at different temperatures: competitive specific binding (% of reference) at three different temperatures.

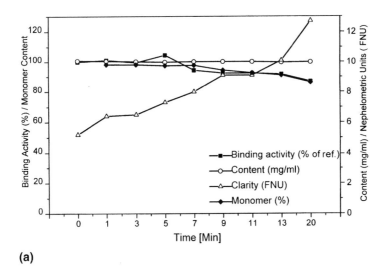

(a)

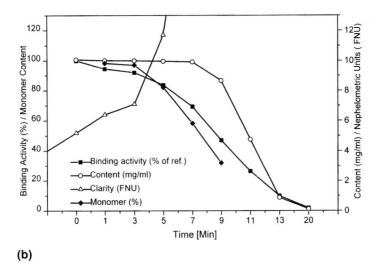

(b)

Fig. 12   Accelerated stability testing of a monoclonal antibody at 65°C.

One of the first goals of early formulation studies is to establish the degradation profile for a protein by accelerated stability testing, employing high temperatures, pH extremes, shear forces due to agitation, aeration, and other stresses. Based on the knowledge of all degradation pathways for a protein, strategies for optimization of the formulation can be applied. Figure 12 illustrates the results of accelerated stability testing for a monoclonal antibody at 65 and 70°C. At both temperatures aggregation is first indicated by the occurrence of opalescence in the solution, followed by other parameters, such as decrease in monomer content (increase of soluble aggregates), decrease in binding activity and, only much later, decrease in protein content due to significant precipitation. This pattern is consistent with the observation that the most common degradation product for many proteins at elevated temperatures is the formation of aggregates. In a study of recombinant t-PA, stability was assessed at 25°C for about one month in a 0.2 M arginine phosphate buffer system at various pH values. The stability-indicating test methods employed were HPLC-SEC for monomer content, HPLC-SEC with detergent in the running buffer for quantitation of two-chain rt-PA, and the clot lysis assay to measure potency. The pH optimum for rt-PA in this formulation was at pH of 6.0; a decrease in monomer content and in single-chain rt-PA occurred at higher pH values, whereas a decrease in clot lysis activity occurred at lower pH values (133).

Analytical methods are applied in the initial screening of formulation candidates that have been identified as indicative for the stability profile of a given protein pharmaceutical, reflecting the major degradation pathways discussed earlier. Based on screening studies with several formulation candidates at accelerated temperatures and on further optimization of formulation compositions, one or two candidates are chosen for long-term real-time stability studies. The overall design and scope of such a stability study as well as parameters to be tested are described in the tripartite guideline of the International Conference on Harmonization (134); an annex to this document deals with the specific requirements for biotechnological/biological products (135).

V. SUMMARY

This chapter describes the analytical methods used to characterize protein drugs and support formulation development and stability testing. The presentation of methods focuses on the criteria an analytical method must meet to be stability indicating and useful in screening formulations and in formal stability programs. A brief discussion of the manufacturing process is provided, be-

cause the process for production of a biopharmaceutical can have a strong
influence on the resulting characteristics and stability profile of the protein.
The importance of producing a consistent product by strict control and validation of fermentation and purification processes is discussed. An extensive
physicochemical characterization of the proteins forms the basis for choosing
specific methods for formulation development, as well as for product release
and stability testing. The stability-indicating methods should detect the most
common degradation forms of protein products. The selection of methods used
for an individual protein will be determined, at least in part, by the properties
of the protein and the design criteria for the formulation. The common degradation pathways of proteins are discussed, with emphasis on analytical techniques useful for monitoring these changes in proteins during stability testing.
The understanding of the principles of these analytical methods, the determination of factors controlling the major degradation pathways, and the selection
of appropriate strategies for minimizing such degradation are prerequisites for
the development of stable formulations.

ACKNOWLEDGMENTS

The analytical data shown as examples in the figures were generated in the
quality control laboratories of Biopharmaceutical Manufacture, Boehringer Ingelheim, Biberach, and were kindly provided by Dr. Michael Schlüter, Dr.
Stefan Bassarab, Dr. Jochen Wallach, and Dr. Kristina Kopp. I also thank my
analytical colleagues for many fruitful discussions on these topics.

REFERENCES

1. R.-G. Werner, Arzneim.-Forsch./Drug Res. 36:1429–1436 (1998).
2. R.-G. Werner, F. Walz, W. Noé, and A. Konrad, J. Biotechnol. 22:51–58
    (1992).
3. R.-G. Werner and C.-H. Pommer, Arzneim.-Forsch./Drug Res. 40(II) 11:1274–
    1283 (1990).
4. R.-G. Werner, Clinical Pharmacology of Biotechnology Products. Amsterdam:
    Elsevier Science Publishers, 1991, pp. 3–22.
5. K. Kopp, M. Schlüter, and R.-G. Werner, In: R. R. Townsend, and A. T. Hotch-
    kiss, Jr., eds. Techniques in Glycobiology. New York: Marcel Dekker, 1997,
    pp. 475–489.
6. H. Allgaier, Chimia 48:464–466 (1994).

7. R.-G. Werner and W. Noé, Arzneim.-Forsch./Drug Res. 43(II) 10:1134–1139 (1993).
8. R.-G. Werner and W. Noé, Arzneim.-Forsch./Drug Res. 43(II) 11:1242–1249 (1993).
9. W. Berthold and J. Walter, Biologicals 22:135–150 (1994).
10. Biotechnology-derived articles. In: The United States Pharmacopeia. Rockville, MD: U.S. Pharmacopeial Convention, Inc., 1994, pp. 1849–1859.
11. A. J. S. Jones, Adv. Drug Delivery Rev. 10:29–90 (1993).
12. K. Kopp, M. Schlüter, and R.-G. Werner, In: R. R. Townsend and A. T. Hotchkiss, Jr., eds. Techniques in Glycobiology. New York: Marcel Dekker, 1997, pp. 475–489.
13. W. S. Hancock, C. A. Bishop, and M. T. W. Hearn, Anal. Biochem. 92:170–173 (1979).
14. R. C. Chloupek, R. J. Harris, C. K. Leonard, R. G. Keck, B. A. Keyt, M. W. Spellman, A. J. S. Jones, and W. S. Hancock, J. Chromatogr. 463:375–396 (1989).
15. J. J. Dougherty, L. M. Snyder, R. L. Sinclair, and R. H. Robbins, Anal. Biochem. 190:7–20 (1990).
16. H. R. Morris, M. Panico, and G. W. Taylor, Biochem. Biophys. Res. Commun. 117:299–305 (1983).
17. B. W. Gibson and K. Biemann, Proc. Natl. Acad. Sci. U.S.A. 81:1956–1960 (1984).
18. R. M. Caprioli, B. DaGue, T. Fan, and W. Moore, Biochem. Biophys. Res. Commun. 46:291–299 (1987).
19. P. Edman and C. Begg, Eur. J. Biochem. 1:80–91 (1967).
20. A. J. S. Jones and R. L. Garnick, In: A. S. Lubniecki, ed., Large-Scale Mammalian Cell Culture Technology. New York: Marcel Dekker, 1990, pp. 543–566.
21. B. W. Gibson, Z. Yu, B. Gillece-Castro, W. Aberth, F. C. Walls, and A. L. Burlingame, In: T. E. Hugli, ed. Techniques in Protein Chemistry, Vol. 3., San Diego: Academic Press, 1989, pp. 135–151.
22. H. A. Havel, R. S. Chao, R. J. Haskel, and T. J. Thamann, Anal. Chem. 61: 642–650 (1989).
23. E. H. Strickland, CRC Crit. Rev. Biochem. 2:113–175.
24. Y.-H. Chen, J. T. Yang, and K. H. Chau, Biochemistry 13:3350–3359 (1974).
25. W. C. Johnson, Annu. Rev. Biophys. Chem. 17:145–166 (1988).
26. R. Pearlman and T. H. Nguyen, Biochemistry 10:824–830 (1991).
27. T. A. Bewley, Recent Prog. Hormone Res. 35:155–213 (1979).
28. A. J. S. Jones and J. V. O'Connor, In: J. L. Gueriguian, E.D., Bransome, Jr., and A. S. Outschoorn, eds., Hormone Drugs. Rockville, MD: U.S. Pharmacopeia, 1982, pp. 335–351.
29. S. R. Davio and M. J. Hageman, In: J. Y. Wang and R. Pearlman, eds., Stability and Characterization of Protein and Peptide Drugs: Case Histories. New York: Plenum Press, 1993, pp. 59–90.

30. H. Susi and D. M. Byler, Methods Enzymol. 130:290–311 (1986).
31. W. K. Surewicz, H. H. Mantsch, and D. Chapman, Biochemistry 32:389–394 (1993).
32. J. L. Arrondo, A. Muga, J. Castresana, and F. M. Goni, Prog. Biophys. Mol. Biol. 59:23–56 (1993).
33. S. J. Prestrelski, T. Arakawa, and J. F. Carpenter, Arch. Biochem. Biophys. 303:465–473 (1993).
34. J. F. Carpenter, S. J. Prestrelski, and T. Arakawa, Arch. Biochem. Biophys. 303: 456–464 (1993).
35. M. J. Kronman and L. G. Holmes, Photochem. Photobiol. 14:113–134 (1971).
36. E. A. Burstein, N. S. Vedenkina, and M. N. Ivkova, Photchem. Photobiol. 18: 263–279 (1973).
37. J. Montreuil, Pure Appl. Chem. 42:431–477 (1975).
38. J. Montreuil, Adv. Carbohydr. Chem. Biochem. 37:157–223 (1980).
39. E. Tsuda, G. Kawanishi, M. Ueda, S. Masuda, and R. Sasaki, Eur. J. Biochem. 188:405–411 (1990).
40. J. Wölle, H. Jansen, L. C. Smith, and L. Chan, J. Lipid Res. 34:2169–2176 (1993).
41. R. B. Parekh, R. A. Dwek, P. M. Rudd, J. R. Thomas, T. W. Rademacher, T. Warren, T.-C. Wun, B. Hebert, B. Reitz, M. Palmier, T. Ramabhadran, and D. C. Tiedmeier, Biochemistry 28:7670–7679 (1989).
42. A. D. Elbein, Trends Biotechnol. 9:346–352 (1991).
43. P. W. Berman and L. A. Lasky, Trends Biotechnol. 3:51–53 (1985).
44. C. F. Steer and G. Ashwell, Prog. Liver Dis. 8:99–123 (1986).
45. M. Takeuchi, N. Inoue, T. W. Strickland, M. Kubota, M. Wada, R. Shimuzu, S. Hoshi, H. Kozutsumi, S. Takasaki, and A. Kobata, Proc. Natl. Acad. Sci. U.S.A. 86:7819–7822 (1989).
46. C. F. Goochee and T. Monica, Biotechnology 8:421–427 (1990).
47. C. F. Goochee, M. J. Gramer, D. C. Anderson, J. B. Bahr, and J. R. Rasmussen, Biotechnology 9:1347–1355 (1991).
48. M. Gawlitzek, H. S. Conradt, and R. Wagner, Biotechnol. Bioeng. 467:536–544 (1995).
49. M. C. Borys, D. I. H. Linzer, and E. T. Papoutsakis, Biotechnol. Bioeng. 43: 505–514 (1994).
50. M. C. Borys, D. I. H. Linzer, and E. T. Papoutsakis, Biotechnology 11:720–724 (1993).
51. M. J. Gramer and C. F. Goochee, Biotechnol. Prog. 9:366–377 (1993).
52. M. J. Gramer, D. V. Schaffer, M. B. Sliwkowski, and C. Goochee, Glycobiology 4:611–616 (1994).
53. M. R. Hardy and R. R. Townsend, Proc. Natl. Acad. Sci. U.S.A. 85:3289–3293 (1988).
54. R. R. Townsend and M. R. Hardy, Glycobiology 1:139–147 (1991).
55. O. H. Lowry, N. J. Rosebrough, A. L. Farr, and R. J. Randall, J. Biol. Chem. 193:265–275 (1951).

56. M. M. Bradford, Anal. Biochem. 72:248–254 (1976).
57. D. B. Wetlaufer, Adv. Protein Chem. 17:303–390 (1962).
58. Y. J. Wang, In: K. E. Avis, H. A. Lieberman and L. Lachman, eds., Pharmaceutical Dosage Forms: Parenteral Medications. New York: Marcel Dekker, 1992, pp. 283–319.
59. A. F. Winder and W. L. G. Gent, Biopolymers 10:1243–1252 (1971).
60. S. C. Gill and P. H. Hippel, Anal. Biochem. 182:319–326 (1989).
61. T. A. Bewly, Anal. Biochem. 123:55–65 (1982).
62. B. D. Hames and D. Rickwood, eds., Gel Electrophoresis of Proteins: A Practical Approach. Oxford: IRL Press, 1981.
63. K. Kleparnik and P. Boocek, J. Chromatogr. 569:3–42 (1991).
64. P. B. Righetti, Isoelectric Focusing: Theory, Methodology and Applications. Amsterdam: Elsevier, 1983.
65. P. G. Righetti, In: T. E. Creighton, ed. Protein Structure: A Practical Approach. Oxford: IRL Press, 1989, pp. 23–63.
66. P. G. Righetti, Immobilized pH Gradients: Theory and Methodology. Amsterdam, Elsevier, 1990.
67. O. Vesterberg and H. Svensson, Acta. Chem. Scand. 20:820–834 (1966).
68. T. H. Nguyen and C. Ward, In: J.-Y. Wang and R. Pearlman, eds. Stability and Characterization of Protein and Peptide Drugs: Case Histories. New York: Plenum Press, 1993, pp. 91–134.
69. E. Canova-Davis, G. M. Teshima, T. J. Kessler, P.-J. Lee, A. W. Guzzetta, and W. S. Hancock, Am. Chem. Soc. Symp. 434:90–112 (1990).
70. R. Pearlman and T. H. Nguyen, In: D. Marshak and D. Liu, eds. Therapeutic Peptides and Proteins: Formulation, Delivery and Targeting. Plainview, NY: Cold Spring Harbor Laboratory, 1989, pp. 23–30.
71. M. P. Henry, In: R. W. A. Oliver, ed. HPLC of Macromolecules: A Practical Approach. Oxford: IRL Press, 1989, pp. 9–125.
72. L. A. Ae. Sluyterman and O. Elgersma, J. Chromatogr. 150:17–30 (1978).
73. L. A. Ae. Sluyterman, Trends Biochem. Sci. 7:168–170 (1982).
74. J. A. Reynolds and C. Tanford, J. Biol. Chem. 245:5161–5165 (1970).
75. Y. P. See and G. Jackowski, In: T. E. Creighton, ed. Protein Structure: A Practical Approach, Oxford: IRL Press, 1989, pp. 1–21.
76. A. L. Shapiro, E. Vinuela, and J. V. Maizel, Biochem. Biophys. Res. Commun. 28:815–820 (1967).
77. K. Weber and M. Osborne, J. Biol. Chem. 244:4406–4412 (1969).
78. U. K. Laemmli, Nature 27:680–685 (1970).
79. D. M. Neville, Jr., J. Biol. Chem. 246:6328–6334 (1971).
80. B. D. Hames, In: B. D. Hames and D. Rickwood, eds. Gel Electrophoresis of Proteins: A Practical Approach. Oxford: IRL Press, 1981, pp. 1–91.
81. K. M. Gooding and F. E. Regnier, In: K. M. Gooding and F. E. Regnier, eds. HPLC of Biological Macromolecules. New York: Marcel Dekker, 1990, pp. 47–75.
82. P. Andrews, Methods Biochem. Anal. 18:1–53 (1970).

83. R. Whitaker, Anal. Chem. 35:1950–1953 (1963).
84. B. M. Van Liedekerke, H. J. Nelis, J. A. Kint, F. W. Vanneste, and A. P. De Leenheer, J. Pharm. Sci. 80:11–16 (1991).
85. M. W. Townsend and P. P. DeLuca, J. Pharm. Sci. 80:63–66 (1991).
86. R. D. Smith, J. A. Loo, R. R. O. Loo, M. Busman, and H. R. Udseth, Mass Spectrom. Rev. 10:359–451 (1991).
87. F. Maquin, B. M. Schoot, P. G. Devaux, and B. N. Green, Rapid Commun. Mass Spectrom. 5:299–302 (1991).
88. F. Hillenkamp and M. Karas, Methods Enzymol. 193:280–294 (1990).
89. M. Karas, U. Bahr, and U. Giessmann, Mass Spectrom. Rev. 10:335–357 (1991).
90. R. C. Beavis and B. T. Chait, Proc. Natl. Acad. Sci. U.S.A. 87:6873–6877 (1990).
91. M. D. Groesbeck and A. F. Parlow, Endocrinology 120:2582 (1987).
92. I. Gery, R. K. Gershon, and B. H. Waksman, J. Exp. Med. 136:128–142 (1972).
93. I. Gery, P. Davies, J. Derr, N. Krett, and J. A. Barranger, Cell. Immunol. 64: 293–303 (1981).
94. L. Gu and J. Fausnaugh, In: Y. J. Wang and R. Pearlman, eds. Pharmaceutical Biotechnology, Vol. 5, Stability and Characterization of Proteins and Peptid Drugs: Case Histories. New York: Plenum Press, 1993, pp. 221–248.
95. W. E. Stewart, II, The Interferon System, Vienna: Springer Verlag, 1981, pp. 13–25.
96. N. B. Finter, Interferon assays and standards. In: N. B. Finger, ed. Interferons. Amsterdam: North Holland, 1966, p. 87.
97. M. Ranby, Biochim. Biophys. Acta 704:461–469 (1982).
98. D. Collen, G. Tytgat, and M. Verstaete, J. Clin. Pathol. 21:705–707 (1968).
99. P. J. Gaffney and A. D. Curtis, Thromb. Haemostasis 53:134–136 (1985).
100. R. H. Carlson, R. L. Garnick, A. J. S. Jones, and A. N. Meunier, Anal. Biochem. 168:28–435 (1988).
101. A. J. Jones and A. M. Meunier, Thromb. Haemostasis 64:455–463 (1990).
102. R. G. Werner, S. Bassarab, H. Hoffman, and M. Schlüter, Arzneim.-Forsch./ Drug Res. 41(II) 11:1196–1200 (1991).
103. V. T. Kung, P. R. Panfili, F. L. Sheldon, R. S. King, P. A. Nagainis, B. Gomez, D. A. Ross, J. Briggs, and R. F. Zuk, Anal. Biochem. 187:220 (1990).
104. L. J. Schiff, W. A. Moore, J. Brown, and H. Wisher, BioPharm 5(5):36–39 (1992).
105. D. Hill and M. Beatrice, BioPharm 2(10):28–32 (1989).
106. R.-G. Werner and J. Walter, BioEngineering 5/90:14–19 (1990).
107. J. Walter and H. Allgaier, In: H. Hauser and R. Wagner, eds. Mammalian Cell Biotechnology in Protein Production. Berlin, New York: Walter de Gruyter, 1997.
108. Q5A Viral Safety Evaluation of Biotechnology Products Derived from Cell Lines of Human or Animal Origin. Fed Regist 63(185):51075–51084 (1998).

109. Q5D Quality of Biotechnological/Biological Products: Derivation and Characterization of Cell Substrates Used for Production of Biotechnological/Biological Products. Fed Regist 63(182):50245–50249 (1998).
110. A. J. S. Jones, In: J. L. Cleland and R. Langer, eds. Formulation and Delivery of Proteins and Peptides. Washington, DC: American Chemical Society, 1994, pp. 22–45.
111. A. K. Banga, In: Therapeutic peptides and proteins: formulation, processing and delivery systems, Lancaster, PA: Technomic Pub, 1995, pp. 61–80.
112. M. C. Manning, K. Patel, and R. T. Borchardt, Pharm. Res. 6:903–918 (1989).
113. B. S. Chang, C. S. Randall, and Y. S. Lee, Pharm. Res. 10:1478–1483 (1993).
114. J. F. Back, D. Oakenfull, and M. B. Smith, Biochemistry 18:5191–5196 (1979).
115. S. Vermuri, I. Beylin, V. Sluzky, P. Stratton, G. Eberlein, and Y. J. Wang, J. Pharm. Pharmacol. 46:481–486 (1994).
116. R. Pearlman and T. Nguyen, J. Pharm. Pharmacol. 44:178–185 (1992).
117. J. L. Cleland, M. F. Powell, and S. J. Shire, Crit. Rev. Thera. Drug Carrier Syst. 10:307–377 (1993).
118. A. F. Winder and W. L. G. Gent, Biopolymers 10:1243–1252 (1971).
119. C. Washington, Particle Size Analysis in Pharmaceutics and Other Industries: Theory and Practice. New York: Ellis Horwood 1992, Chap. 7.
120. D. F. Nicoli and G. B. Benedek, Biopolymers 15:2421–2437 (1976).
121. J. L. Cleland and D. I. C. Wang, Biochemistry 29:11072–11078 (1990).
122. J. Geigert, J. Parenter. Sci. Technol. 43(5):220–224 (1989).
123. J. Wallach, M. Schlüter, and R. G. Werner, Stable formulations for biopharmaceuticals. In: D. Duchene, ed. Recent Advances In Peptide and Protein Delivery, Minutes of the 8th International Pharmaceutical Technology Symposium. 1997.
124. B. R. Oakley, D. R. Kirsch, and N. R. Morris, Anal. Biochem. 105:361–363 (1980).
125. C. R. Merril, D. Goldman, S. A. Sedman, and M. H. Ebert, Science 211:1437–1438 (1981).
126. J. H. Morrissey, Anal. Biochem. 117:307–310 (1981).
127. H.-M. Poehling and V. Neuhoff, Electrophoresis 2:141–147 (1981).
128. J. A. Schrier, R. A. Kenley, R. Williams, R. J. Corcoran, Y. K. Kim, R. P. Northey, D. Daugusta and M. Huberty, Pharm. Res. 10:933–944 (1993).
129. R. A. Kenley and N. W. Warne, Pharm. Res. 11(1):72–76 (1994).
130. M. F. Powell, In: J. L. Cleland and R. Langer eds. Formulation and Delivery of Proteins and Peptides. Washington, DC: American Chemical Society, 1994, pp. 100–117.
131. T. Arakawa, S. J. Prestrelski, W. C. Kenney, and J. F. Carpenter, Adv. Drug Delivery Rev. 10:1–28 (1993).
132. ICH Draft Guideline on the Validation of Analytical Procedures. Fed. Regist. FDA, March 1996.
133. H. Hoffmann, S. Bassarab, M. Schlüter, W. Werz, and R.-G. Werner, Stability Testing of Biopharmaceuticals. In: Paperback APV, Band 32, Stability Testing

in the EC, Japan and the USA, Stuttgart: Wissenschaftliche Verlag, 1993, pp. 245–272.

134.   ICH Guidance, Stability Testing of New Drug Substances and Products, Fed. Regist. 59:48754 (1994).

135.   ICH Guidance, Stability Testing of Biotechnological/Biological Products, FDA Docket No. 93d-01391, Washington, DC: Government Printing Office, 1996.

4

The Importance of a Thorough Preformulation Study

Eugene J. McNally and Christopher E. Lockwood
Boehringer Ingelheim Pharmaceuticals, Inc.
Ridgefield, Connecticut

I. INTRODUCTION

To develop a rational strategy for the selection of stable protein formulations, one needs to have an understanding of the underlying mechanisms of protein instability and degradation, and the analytical techniques available to study these processes. Each of these issues has been covered at length in the preceding two chapters. The distinction between preformulation and formulation experiments is often not consistent from institution to institution. For the purpose of this chapter, preformulation encompasses the phase of formulation development in which the pharmaceutical scientist takes an initial ''look'' at the protein molecule to identify the conditions that are likely to be best suited for development of a formulation possessing long-term stability. With these conditions in hand, a rational decision on formulation conditions that are best suited for the stability of the molecule, the route of delivery, compatibility with the container/device, and in vivo stability can be explored in the formulation phase of development.

Information obtained at this stage of the formulation/development process gives indications that the molecule is potentially degrading, though often the exact mechanism of degradation will not become apparent until much further investigation has been conducted with analytical methods that tend to be highly product specific and often are not in place during the preformulation

111

stage of product development. It is the intent at this stage of development to get a feel for the conditions that cause degradation, as manifested by changes in the molecular weight, charge, and activity of the molecule. This phase of the study will also yield important information on what excipients, stabilizers, and other components might need to be examined to develop a stable formulation.

While preformulation can be considered to be an entity independent of the development of the manufacturing process for production of the drug substance, our experience has shown that preformulation is an integral part of the entire development process. Many of the degradation processes discussed previously and their underlying causes can occur at many stages of the production process starting with cell culture, during downstream processing to recover and purify the desired product, during final dosage form manufacture and shelf storage of the product, and ultimately during in vivo use of the product.

In addition to conditions the protein is exposed to during drug substance manufacture, it is important to begin anticipating the various environments the molecule will be exposed to during final product manufacture. Will the final dosage form be a solution or a lyophilized powder? Will the protein be formulated in a controlled release dosage form? Is the protein being formulated to be delivered to the lung via a delivery device? Is the protein being conjugated to other materials to modify its properties [imaging labels, cytotoxins, poly(ethylene) glycol, targeting agents etc.]? All these product strategies will give rise to different environments that can lead to possible degradation of the protein. Consideration of these various environments should be incorporated into the design of accelerated stability studies, as discussed at the end of this chapter. Table 1 outlines some of the underlying causes of protein instability.

Table 1 Underlying Causes of Instability Encountered During Drug Substance and Drug Product Manufacturing

pH extremes	Radiation
Ionic strength	Residual moisture
Freeze/thaw effects	Shear forces
Organic solvents	Adsorption at interfaces
Light	Temperature
Oxygen	Heavy metal ions
Surfactants	

Key decisions in design of cell culture and purification processes are made early in product development. It is highly desirable to embark on the preformulation efforts early in product development to gain an understanding of the conditions under which instability will arise. Ideally, the preformulation process should start as soon as there are sufficient quantities of drug substance (as little as tens of milligrams), which is often during the end of the drug discovery process. Initiation of these studies early in the development process makes much of the information collected about how to handle the molecule useful to the cell culture and purification development groups. In addition, early studies can also lead to valuable information that may explain changes seen in the molecule with modifications in the manufacturing process during process optimization and scale-up.

The exact timing for the initiation of preformulation studies needs to be balanced with the availability of material that is likely to be representative of the product used in the preclinical toxicology and early clinical trials. The use of early "discovery" material in preformulation studies can be problematic owing to differences in the purity of the molecule, the mode of production, and the methods used for product recovery. These differences can have dramatic effects on the physical and chemical characteristics of the molecule and its ultimate stability profile.

Regardless of the timing for initiation of these studies, it is imperative that these preformulation studies be performed before the excipients and storage conditions that constitute the formulation are decided. As Fig. 1 depicts, scientists from many different disciplines of the discovery, development, and manufacturing process wish to participate in selection of the formulation. However, the importance of performing thorough preformulation and formulation studies is not always realized until it is too late, and the drug product fails release testing or is found to have too short a shelf-life to be useful in clinical investigations or as an ultimate commercial product. The consequences of a poorly performed preformulation study, or no study, are increased time and cost of product development, insufficient product stability, the need to register a suboptimal formulation or restrictive storage conditions, relative to a competitor's product.

The primary aim of a preformulation study is to determine the inherent stability of the molecule (drug substance) and to identify the key problems that are likely to be encountered in development of a stable formulation. A typical preformulation study can include the examination of a variety of characteristics, some of which are listed in Table 2 (1). This chapter addresses the characteristics that pertain to proteins in solution, since this is typically the state of most protein drug substances early in development. If a lyophilized

Fig. 1 Everyone wants to formulate the product. No one wants to do preformulation. (Illustrated by Leigh Rondano, Boehringer Ingelheim Pharmaceuticals, Inc.)

formulation is desired, the appropriate preformulation characteristics specific to such solids should be evaluated during the initial phase of lyophilization development.

The key components of a preformulation study can be broken down into a series of ordered activities. The first objective is to establish analytical methods and use these techniques to characterize the chemical and physical proper-

Table 2 Potential Characteristics of a Preformulation Profile

Structure	Identification of key degradations
Molecular weight	Melting point
pH–solubility profile	Absorbance spectra
pH–conformational changes	Solvate formation
Air–water interface effects	Polymorphism potential
Effects of freezing and thawing	Hygroscopicity potential
Organic solvent compatibility	Accelerated stability (effects of time, light, temperature, oxygen)

Source: Ref. 1.

ties of the purified drug substance. This first step is essentially an initial "look" at the molecule. The second phase of a preformulation study is determination of the reactivity of the molecule as a function of conditions likely to be encountered during processing and final formulation. Elucidation of the reactivity of the molecule with respect to pH, ionic strength, temperature, and concentration will enable rational decisions to be made when it is time to consider suitable excipients and components for use in formulation studies. While this examination will not discern all potential degradation mechanisms, it will provide information about proper handling and storage conditions, and it can be used to rule out potential formulations likely to be unstable. The last phase of preformulation is the identification of the key degradation mechanisms that could be encountered during drug product stability testing. This requires the isolation of degradation products or identification of reaction by-products. Often more sophisticated analytical methods, which are not generally available early in product development, are required to identify the degradation products. This third stage of preformulation is often an ongoing activity and continues into the later stages of product development.

II. INITIAL CHARACTERIZATION OF THE MOLECULE

The first phase of preformulation serves as an initial "look" at the properties of the protein molecule. For example, information on the molecular weight, isoelectric point, aggregation state, hydrophobicity, solubility, and melting temperature, as well as any information on the inherent molecular heterogeneity of the product, is of particular importance. To assess these characteristics, one must first establish appropriate analytical techniques.

A. Analytical Methods Used in Preformulation

Because of the complexity of protein molecules, assessment of changes in the molecule and stability must be performed with a battery of analytical methods. Unlike conventional, small, synthetic molecules, which can be completely characterized in terms of primary structure, one needs to consider the effects of chemical and physical degradation on primary, secondary, and tertiary protein structure. However, at this early stage of development, most of the instabilities listed in Table 3 can be detected by monitoring the protein for changes in charge, size and aggregation state, visible appearance, and functional activity (2).

Table 3   Major Degradation Pathways Leading to
Protein Instability

Hydrolysis	Deamidation
Denaturation/Aggregation	Disulfide scrambling
Oxidation	Deglycation

Source: Ref. 2.

While monitoring of these attributes will not yield exact information on the underlying causes of instability, it will allow one to rule out certain conditions as being unacceptable. Several standard analytical methods can often be applied to track these changes for a particular compound. These four characteristics, the associated assays used to track them, and examples of typical data are in the subsections that follow.

1. Charge

Isoelectric focusing (IEF) measures the isoelectric point of the molecule and indicates any changes in the overall net charge of the protein. Recall that the isoelectric point is defined as the pH at which the net charge on the molecule is zero. Figure 2 shows two IEF gels for a murine monoclonal antibody that was stored under a variety of pH conditions. In this IEF method, when current is applied to the gel, a pH gradient is established, and the protein migrates and focuses at its isoelectric point.

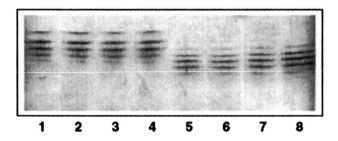

Fig. 2   A Coomassie blue stained pH 3–9 isoelectric focusing gel for a murine monoclonal antibody stored for 8 days at 25°C. Protein was at a concentration of 1 mg/mL in a 50 mM phosphate, 100 mM sodium chloride buffer pH 7.5 (lanes 1–4), and a 50 mM glycine, 100 mM sodium chloride buffer pH 9.0 (lanes 5–8).

2. Size

Size can be assessed by both electrophoretic and chromatographic techniques, which give different information about the state of the molecule. A commonly used HPLC technique, size exclusion chromatography (SEC), separates molecules based on their hydrodynamic volume. The largest materials elute prior to smaller components. Figure 3 shows a chromatogram for a recombinant glycoprotein that tends to aggregate. The aggregate is shown to be consistent with a dimer that reverts to monomer when analyzed by a denaturing electrophoretic technique, sodium dodecyl sulfate–polyacrylamide gel electrophoresis (SDS-PAGE), run under reducing conditions (see Fig. 4). The use of both these techniques together provides information on the native and denatured

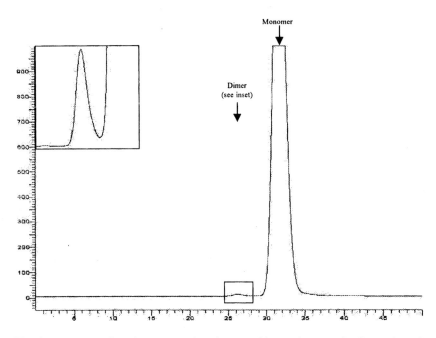

Fig. 3 Size exclusion chromatography of a recombinant glycoprotein. Separation of monomer and dimer species was accomplished using two Pharmacia Superdex 200 HR 10/30 columns in series (column length 300 mm, particle size 15 µm, column diameter 10 mm). The columns were eluted with 0.75 mL/min 50 mM sodium phosphate/125 mM NaCl, pH 7.0 at room temperature. Protein elution was monitored using fluorescence detection at excitation and emission wavelengths of 285 and 343 nm, respectively.

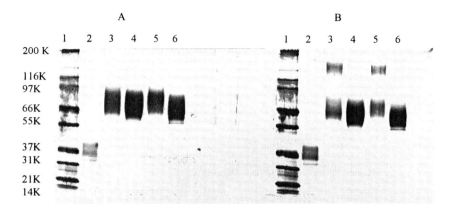

Fig. 4 SDS-PAGE gradient gel (4–12% Tris–glycine) for a recombinant glycoprotein run under reduced (gel A) and nonreduced (gel B) conditions and developed with silver staining. Samples were generated by taking fractions from a protein sample eluting off of a size exclusion chromatography column (see Fig. 3 for conditions): lane 1, molecular weight marker proteins; lane 2, molecular weight marker protein; lanes 3 and 5, aggregate fractions; lanes 4 and 6, monomer fractions.

states of the molecule, the covalent/noncovalent nature of protein aggregates, and also on the presence of smaller molecular weight protein variants/degradation products that might be evident under the denaturing conditions of the SDS-PAGE gel but not under the native nondenaturing conditions of SEC. By definition, soluble aggregates are species that pass through and can be detected on size exclusion chromatography columns. Insoluble aggregates are those that can be removed from solution by physical means (e.g., filtration or centrifugation).

Another electrophoretic technique that separates proteins based on both size and charge is native-PAGE. Native-PAGE can be useful for detecting small traces of soluble and insoluble protein aggregates that may never enter the SEC column or be detectable under the denaturing conditions of SDS. On a native-PAGE gel, insoluble aggregates can often be identified as species that are too large to migrate into the gel and precipitate at the point of application on the gel. Another advantage of native-PAGE over SEC is the ability to assess the protein in its native conditions. A limitation of SEC is the need to include salt in the mobile phase to minimize protein interactions with the column. The high ionic strength of the mobile phase can disturb weak interactions associated with some noncovalent oligomeric species and give rise to artifacts that would not arise with native-PAGE (2). However, not all proteins lend themselves to native-PAGE analysis. Proteins that possess a large degree of charge heterogeneity will not always focus sufficiently on native-PAGE gels to make quantitation of the gels possible. Ultimately, the best approach is to use a combination of all these techniques to assess the size and aggregation state of a protein molecule.

3. Visual Appearance

The observation of a product's visual appearance, while simplistic in nature, is often overlooked for its value in detecting changes in a molecule. The range of this assay can vary from a qualitative description of color and clarity of the solution all the way to a quantitative method such as monitoring the solution for changes in optical density (3).

4. Functional Activity

The biological activity assay demonstrates the functional activity of the molecule. Unlike traditional small molecular weight chemical entities, biologics are highly complex molecules that possess primary, secondary, and tertiary structure. Because of the need for short development timelines, the complete analytical characterization of the chemical and physical state of these molecules by techniques such as nuclear Overhauser enhancement spectroscopy (NOESY) and various X-ray spectroscopies is not practical. However, subtle changes in the chemical and physical structure of the protein can lead to a loss of activity; therefore, the assessment of biological activity is an integral part of the formulation development program. These activity assays are specific to a product and its intended in vivo indication (i.e., functional assays) and are often not of the most sensitive nature. In addition, their adaptation to a quantitative analytical assay format is a time-consuming and difficult process, which usually occurs in parallel to initial preformulation/formulation work. Our approach has been to initiate preformulation work using the above-described assays that assess changes in the molecule's charge, size/aggregation state, and visible appearance (2). Once potential conditions for handling the molecule have been identified and a suitable functional activity assay has been established, one can go back and investigate whether the results of the initial assays correlate with changes in functional activity.

5. Qualification of the Method

Validated stability-indicating methods are not usually available at this early stage of preformulation; however, it is advisable to demonstrate that the

Table 4 Example of Stressed Conditions Used During Preformulation Method
Development

Stress[a]	Conditions applied to sample[b]
Acid–base hydrolysis	0.1 M sodium chloride, pH 2.5 (hydrochloric acid) or pH 12.0 (sodium hydroxide), 25°C, for 1 h, pH then readjusted to 6.0
Thermal denaturation	0.1 M sodium chloride, 0.05 M sodium phosphate, pH 6.0, 65°C for 45 min
Disulfide isomerization	0.1 M sodium chloride, 0.05 M sodium phosphate, pH 8.5–9.0, 0.5% mercaptoethanol, 25°C for 3 h; then oxygen bubbled through sample for 3 min
Oxidation	0.1 M sodium chloride, 0.05 M sodium phosphate, pH 6.0, 3% hydrogen peroxide at 25°C for 3 h

[a] Stresses were applied to assess the ability of protein analytical methods to detect a change in a murine monoclonal antibody.
[b] Protein concentration was in the range of 0.2–0.6 mg/mL.

method is capable of detecting changes in the molecule. This usually requires a small study in which the protein is stressed under a variety of conditions, with detection of the corresponding changes that occur. An example of conditions used to stress a molecule for the purpose of evaluating the usefulness of various analytical methods is given in Table 4. Validation of the stability-indicating nature of an analytical method is being demonstrated in parallel to the initial preformulation studies and is a process that continues throughout development of the molecule.

B. Physical and Chemical Characteristics of the Drug Substance

Prior to initiating a screen for potential instabilities, it is helpful to characterize the protein with respect to size and aggregation state of the molecule, isoelectric point, solubility, and melting temperature. In addition to being useful in setting the conditions of the study, these four physical/chemical parameters will be used to monitor and explain changes in the molecule observed under stressed conditions.

1. Size/Aggregation State of the Molecule

It is important to consider the native state of the molecule with respect to its aggregation state. Does the protein exist as a monomer, a homo- or hetero-dimer, or in a variety of monomeric and oligomeric states? Does the molecule possess disulfide bonds? Is the molecule an antibody that possesses heavy and light chains? What information is available regarding the covalent or noncovalent nature of possible oligomeric species? Questions of all these types should be considered when one is first developing analytical methods to evaluate the molecular weight/size of the protein. By using a combination of the electrophoretic and chromatographic techniques discussed above, one can fully appreciate such questions and their impact on what is observed in the stress studies.

2. Isoelectric Point

The isoelectric point can be determined experimentally by either electrophoresis or ion exchange. During preformulation, a method is required that is capable of segregating out subtle changes in the charge pattern over time. In the typical IEF pattern for a murine monoclonal antibody earlier shown (Fig. 2), the changes on the gel are consistent with protein deamidation in which the neutrally charged amide-containing amino acids asparagine and glutamine are converted to negatively charged aspartic acid and glutamic acid, respectively. This process results in a shift of the protein's isoelectric point to a lower pH range. Such results are only suggestive of a deamidation mechanism, and other more definitive analytical methods such as tryptic mapping or methods specific for deamidation (4,5), would need to be performed to rule out other changes in the protein molecule. In addition to containing a variety of charged amino acids, many therapeutic proteins contain charged sugar moieties, such as sialic acid, that will contribute to the IEF pattern. It is often instructive to compare the calculated isoelectric point based on the amino acid sequence alone with the experimental values determined by isoelectric focusing. This calculation can be performed from the amino acid composition of the protein and estimates of the pK_a values of the charged amino acids, including values for the terminal amino and carboxylic acid groups. The charge for each amino acid at a given pH may be calculated and summed over the entire protein sequence to calculate the overall net charge as a function of pH (6). A spreadsheet or simple computer program can be used to find the pH that produces a net charge of zero, or the isoelectric point.

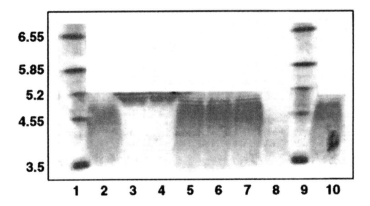

Fig. 5 Isoelectric focusing gel for a highly glycosylated recombinant protein before and after treatment with neuraminidase.

Figure 5 presents an isoelectric focusing gel for a highly glycosylated recombinant protein. The calculated isoelectric point for this molecule is 5.8 when only the ionizable amino acids and the terminal amino and carboxylic acid groups are considered. The isoelectric focusing pattern for the recombinant glycoprotein protein shown in Fig. 5 is extremely complex and heterogeneous. Based on the isoelectric focusing markers in lanes 1 and 9 the pI for this protein is in the range of 3.5–5.2. Lanes 2 and 3 represent the protein after it has been treated with an insoluble form of the enzyme neuraminidase, which cleaves sialic acid groups from the carbohydrate portion of the glycoprotein. The resulting pattern of bands for the enzyme-treated sample yield a pI of approximately 5.2, which is in much closer agreement with the calculated isoelectric point of 5.8, based on the amino acid sequence alone.

3. Solubility

Solubility is an important parameter that may affect the upper concentration limit of a formulated product. It is important to understand the concentration limit of a protein early in development so that appropriate handling conditions can be established. Solubility becomes increasingly important when one is deciding on the pH of a formulation, since many products possess a solubility minimum at the isoelectric point, due to the lack of a net charge on the molecule, profiles of pH solubility and salt solubility are usually generated during formulation development when the excipient composition is being optimized; these studies are described in Chap. 5.

A quick method for assessing protein solubility is ultrafiltration. A membrane with the appropriate molecular weight cutoff is used to concentrate the product such that the product is retained and buffer/solvent is allowed to pass through the membrane. For example, the solubility of a recombinant glycoprotein was determined by concentrating the product in a 30,000 Da Millipore Ultrafree PFL ultrafiltration device. Size exclusion chromatography served as the method of protein quantitation. A sample of protein at 12 mg/mL was concentrated over the course of 10 hours at 4°C. No aggregates were observed in the retentate or in the filtrate, either visually or by SEC, and analysis yielded a concentration of 110 mg/mL. This analytical result represented a concentration limited by the viscosity of the sample, not a true solubility. However, the technique allows for quick estimation of whether solubility will limit the dose used in a formulation, in addition to providing information on upper limits of concentration that can be used during downstream processing of the protein. Solubility can also be determined by lyophilizing the protein in a volatile buffer and then performing saturated solubility measurements of the reconstituted material in the desired buffer system. A further discussion of solubility measurement and techniques for altering protein solubility can be found in Chap. 5.

4. Melting Temperature

The melting temperature of a protein provides one measure of the physical stability of the molecule. Melting temperature (T_m) is defined as the temperature at which equal amounts of native and denatured protein exist in equilibrium. Differential scanning calorimetry (DSC) is a method commonly used to determine the temperature at which the heat-induced unfolding of a protein molecule occurs, although spectroscopic methods are also available (7,8). Conditions producing increases in the T_m (or, more precisely, the Gibbs free energy of denaturation) for a particular protein provide for greater resistance to thermal denaturation and thus greater physical stability. Excipients producing an increase in T_m, therefore, have ordinarily been found to increase physical stability, while those decreasing T_m have been found to decrease physical stability (8–12). A comparison of T_m values obtained for a protein under varying solution conditions, such as the presence or absence of a particular excipient, can then be used as one method for assessing excipient compatibility and the probability of long-term protein stability in a particular formulation. Formulations found to produce large decreases in the melting temperature of a protein can be screened out early in the development process before long-term stability experiments have begun.

Determination of the melting temperature for a particular protein is also important for the design of accelerated stability studies. Studies conducted at temperatures elevated above the T_m may not be indicative of stability under expected storage conditions, since gross changes in higher order structure may alter the primary degradation pathway (13). It should also be noted that some proteins may undergo aggregation or even precipitation upon thermal denaturation, thus limiting the usefulness of DSC (14). In the past, the value of DSC in preformulation studies also was restricted by the need for substantial amounts of protein at a time in development when only limited quantities of material were likely to be available. In recent years, the development of low volume instruments with higher sensitivities has dramatically reduced the amount of protein needed for DSC thermograms (7).

It should also be noted that, as with any method for evaluating protein formulations, DSC results should not be interpreted alone. An increase or decrease in melting temperature should not be used as the sole indicator for the determination of whether to further evaluate a particular formulation or formulation component. Rather, information obtained from DSC studies should be evaluated in conjunction with other preformulation analytical results to choose formulations that merit further development.

III. DETERMINATION OF DRUG SUBSTANCE REACTIVITY

The starting point for preformulation work is the purified bulk drug substance. If stability problems exist for drug substance, these will have to be dealt with in development of dosage forms with long term stability, which may be formulated with other objectives in mind (e.g., controlled delivery).

Our approach is to set up studies of several types lasting from several weeks to months, mimicking conditions known to give rise to instabilities, and then analyze the protein for changes in size/aggregation state, charge, and functional activity. Prior to initiating a reactivity study, it is useful to anticipate the results by considering the protein's primary amino acid sequence. For example, from the study of peptides, protein deamidation is known to preferentially occur at asparagine residues rather than glutamine, and the rate of reaction is known to be greatly enhanced if a glycine amino acid side chain is in the adjacent position in the primary amino acid sequence (15). Ignoring possible contributions of protein tertiary structure, it is useful to examine the primary amino acid sequence and determine how many of these "highly reactive" asparagine/glycine combinations occur in the protein of interest.

Another example is consideration of the number of cysteine amino acid residues a protein possesses. Does the primary amino acid sequence for the molecule possess an odd number of cysteines, which could lead to disulfide scrambling reactions? Have the cystines, which are involved in disulfide bonds, been assigned? Almost always, such studies have not been performed; however, literature sometimes exists in which disulfide bonds have been speculatively assigned on the basis of comparisons with other structurally and functionally related proteins (16). Paper exercises such as these will give hints of potential instabilities that one might encounter and should certainly be on the lookout for.

Conditions explored can range from pH and temperature to the effect of interfaces, organic solvents, freeze/thaw exposure, and any other conditions the molecule will likely be exposed to during processing, formulation, and delivery. On completion of these accelerated studies, any relevant changes observed can be further studied in an attempt to identify the nature of the instability or degradation. When studying proteins, the formulator must be careful to choose accelerated conditions appropriate for the molecule being examined. The choice of accelerated conditions is difficult for proteins because they tend to undergo changes in conformation with temperature. By exceeding the melting temperature or glass transition temperature, the nature of degradation mechanisms is likely to change at and above these critical temperatures. A useful guideline is to stay at least 5–10 degrees below the melting temperature of the protein in solution, or the glass transition temperature of a lyophilized solid, to avoid these changes in higher order structure (2).

IV. IDENTIFICATION OF KEY DEGRADATION PRODUCTS

At this phase of development, the primary goal is not exact elucidation of the degradation mechanisms but rather an assessment of whether the drug substance is stable or is likely to possess major instabilities during both long-term storage in a final dosage form and during processing. While not absolutely necessary, it is often instructive to identify the type of degradation products being generated so that the formulator has information on the types of instability that will be encountered during later stages of the formulation development process. If the cause of instabilities can be determined, one can rationally design formulations under the correct conditions and with appropriate excipients to minimize degradation (13). This rational design approach is much more rewarding and challenging than the factorial approach, which is labor intensive because multiple formulations with a large number of excipient combinations must be examined under a variety of storage conditions. This approach requires considerable analytical testing resources.

In addition, drug supply is often limited at this phase of development, so only a few conditions can be studied. The identification of the impurities almost always requires the use of analytical techniques that are more sophisticated and protein specific than the more general macroscopic methods described in the beginning of this chapter. Chapter 3, on analytical methods, discusses many of these more advanced techniques in detail.

V. PREFORMULATION CASE STUDIES

The following case studies illustrate some of the more interesting changes and degradation processes observed during preformulation studies on a variety of protein molecules. Specifically, we present case studies exploring the effect of pH and temperature extremes on short-term storage and the influence of prolonged shaking and subsequent exposure to the potentially denaturing effects of the air–water interface. The effects of freeze/thaw cycling and the results of short-term compatibility studies with organic solvents are discussed. The use of DSC to screen preservatives likely to be compatible with a protein prior to initiating prolonged and extensive real-time formulation studies is presented. The final case study illustrates the identification of a key degradation mechanism for a recombinant protein and the importance of restarting the preformulation process when changes are made in a formulation or delivery strategy.

A. Effect of pH, Temperature, and Protein Concentration on Short-Term Stability

To assess the chemical stability of a murine monoclonal antibody under conditions relevant to product formulation, the influence on short-term stability of pH (4.5–9.0), at a set concentration of 1 mg/mL and 25°C, was examined. Specifically, the pH conditions examined were: pH 4.5, 50 mM citrate, 100 mM sodium chloride; pH 6.0, 50 mM phosphate, 100 mM sodium chloride; pH 7.5, 50 mM phosphate, 100 mM sodium chloride; and pH 9.0, 50 mM glycine, 100 mM sodium chloride. All solutions were prepared under aseptic conditions and contained no preservatives. Changes in the molecule were monitored by means of a combination of analytical techniques for observing the properties of size and charge: size exclusion chromatography, cation exchange chromatography, native-PAGE, and isoelectric focusing. The predominant changes observed during the 29-day study were effects due to pH on the charge distribution of the molecule. Depending on the conditions, significant shifts in the IEF and native-PAGE electrophoresis patterns were observed for this molecule.

A representative IEF gel from this phase was shown in Fig. 2. To quantitate the changes observed on the electrophoresis gels, a laser densitometer was used to scan the lanes. The area in each band for a given lane was plotted as a function of the calibrated pI. Figure 6 is a graphic representation of the IEF gels, the results of which suggest that the protein is relatively stable with regard to changes in charge distribution up to about pH 6.0. The molecule starts to exhibit significant change in charge at the pH 7.5 condition, and even

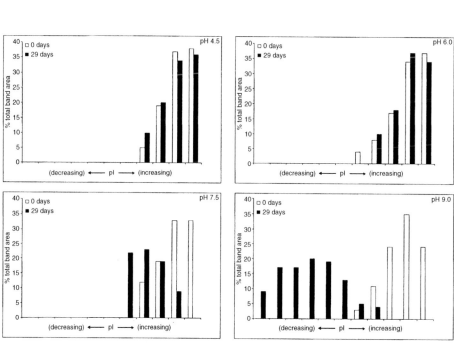

Fig. 6 Quantitative representation of laser densitometric scans of isoelectric focusing gels for a murine monoclonal antibody stored at 25°C at various pH conditions (see text for buffer compositions). Each plot compares the IEF pattern at day zero and day 29 of the study for a given pH condition.

more pronounced shifts to a lower isoelectric point at pH 9.0 over the 29 days studied.

As discussed earlier, shifts in an IEF pattern to lower pH values is suggestive of protein deamidation. A more thorough study in the pH range of 6.0–7.5 would be required to identify the optimum pH to minimize this change in the isoelectric point of the molecule. Such a study is best performed as a part of the formulation phase of product development when specific buffers and other excipients are being examined. The results from the native-PAGE gels (not shown) exhibited the same trend as a function of pH seen in the IEF gels. Analysis of these samples by SEC detected no loss in total soluble protein nor any change in the monomer/aggregate content. No insoluble protein aggregates were evident on the native-PAGE gels. Cation exchange chromatographic analysis revealed significant shifts in the retention time of the protein which were also consistent with changes in the charge distribution of the molecule as a function of increasing pH.

Additional conditions to explore the effect of temperature and protein concentration were explored, with minimal changes observed. The protein showed less change at the lower storage temperature (4°C) than those results at 25°C (Fig. 6), which would be expected for mechanisms such as deamidation. Over the relatively short time period studied, no enhanced stability was observed for material at protein concentrations up to 20 mg/mL.

B. Effect of Shaking on Protein Stability

It has been known for many years that proteins are susceptible to aggregation and precipitation upon shaking (17). It is generally accepted that the pathway leading to aggregate formation is related to the conformational changes a protein molecule undergoes when exposed to an interface, which for most solution formulations is the air–water interface (17). To gauge the degree of susceptibility a protein has for denaturation at the air–water interface, it is useful to perform a short-term cavitation study. A typical experiment will involve filling protein solution at a concentration of approximately 1 mg/mL into glass ampules or serum vials sealed with Teflon-coated gray butyl rubber stoppers. Vials or ampules are typically placed into two configurations (horizontal and vertical) and at various intervals of shaking at 4 or 25°C.

A variety of analytical techniques have been used to quantitate the extent of aggregation. SEC has been used to judge the degree of soluble aggregate formation and to monitor loss of protein content from solution. Native-PAGE is also useful in detecting insoluble aggregate formation. The degree of insoluble aggregate formation is assessed by visual inspection where results can be expressed on a qualitative scale or can be quantitated by turbidity (17), by light scattering, or by measuring changes in the optical density (3,18).

Our general experience with studies of the foregoing types has been that subjecting proteins to vigorous shaking at room temperature can lead to increases in turbidity and insoluble aggregate content with concomitant loss of total soluble protein. Our general observations have been that aggregation is minimal in the vertical configuration or when the vial is completely filled, leaving no headspace. Vial orientation and fill volume both dictate the amount of air–liquid interface available for the protein denaturation and are important variables that can be manipulated to minimize this form of degradation. A number of proteins ranging from recombinant glycoproteins to murine monoclonal antibodies and pegylated antibodies have been examined by means of this technique with mixed results. The greatest degree of aggregation was observed with a monoclonal antibody that resulted in the formation of only insoluble aggregates, which accounted for a 6% loss in total protein content after 7 days of shaking at 25°C. When modified by a covalent pegylation process, the same antibody exhibited significantly less insoluble aggregate formation. Both these proteins showed significantly less aggregation when the experiment was performed at 4°C. This is in contrast to a recombinant protein, which exhibited no degree of aggregation upon shaking for 7 days at room temperature.

Although the results just described are qualitative because of the problem of quantitating the extent and nature of agitation, they help the formulation scientist to determine whether excipients that minimize surface denaturation/aggregation will be needed in a formulation. While shown to denature under conditions of constant shaking at room temperature, these particular proteins are actually quite stable when handled under ordinary conditions. These two proteins serve as in-house references to the severity of denaturation upon shaking and provide a useful link to known behavior under more realistic handling conditions. In the absence of such references, one could make comparisons to commercially available proteins whose behavior on shaking has been studied (17). Later in development, when sufficient quantities of protein are available and a candidate formulation has been chosen, it is important to perform a "shipping test" to mimic normal handling and environmental conditions to which a product will be exposed.

C. Freeze/Thaw Effects

At the preformulation phase of development, it is useful to have an understanding of the effect of the freeze/thaw process on the stability of the protein. At early stages of development, many proteins are handled as unpreserved solutions that often are not prepared in an aseptic environment, so storage at low temperature (freezing) is required to limit microbial growth and minimize other degradative processes.

The process of freezing protein solutions, however, can cause protein aggregation or denaturation. Ice crystals formed during the freezing process may lead to increased concentrations of protein or other formulation constit-

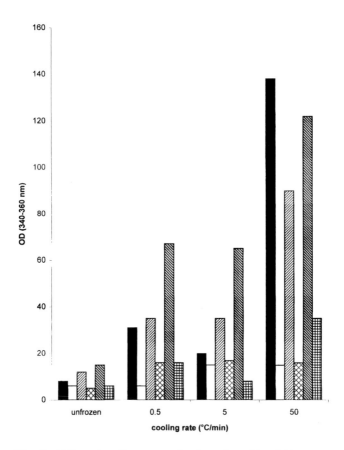

Fig. 7 Effect of cooling rate on the formation of insoluble aggregates of hGH. Bars from left (solid) to right (open squares) as follows: 5 mM phosphate, pH 7.4; 5 mM phosphate, pH 7.8; 88 mM mannitol, 5 mM phosphate, pH 7.4; 88 mM mannitol, 5 mM phosphate, pH 7.8; 250 mM mannitol, 5 mM phosphate, pH 7.8 (control). (Data from Ref. 14.)

uents, which, in turn, can give rise to local extremes of concentration, pH, or ionic strengths. At this stage in development, consideration should also be given to the effect of the rate of freezing. Fast freezing in liquid nitrogen or dry ice–ethanol baths has been reported to minimize the above-mentioned detrimental effects of freezing on some proteins (19,20), while in other cases, quick freezing has been shown to produce higher levels of protein aggregation (3,21,22).

Figure 7 shows data from the work of Eckhardt et al., who studied the effect of cooling rate on the aggregation of human growth hormone (hGH) (3). Figure 7 shows the effect of freezing rates from 0.5 to 50°C/h on the mean optical density of five different formulations of hGH. Higher optical densities are indicative of higher levels of insoluble aggregates. Increasing rates of freezing were associated with an increase in the amount of insoluble aggregates, and a corresponding decrease in soluble hGH monomer. The authors suggest that the large solid–liquid interface, which develops during fast freezing, could be responsible for surface denaturation and subsequent protein aggregation. Similar results have been reported by Chang et al., who found protein aggregation for six different proteins to be highest under fast freezing conditions and to be greatly reduced in the presence of low molecular weight surfactants (22).

These studies illustrate the need for preformulation studies to determine the best freeze/thaw conditions for each protein on a case-by-case basis. To support freezing of purified bulk drug products or final drug products, experiments should be performed at the scale representative of the packaging container intended for use. However, at the preformulation stage of development, the experiment can be scaled down to 0.5–1 mL to conserve material. Later in development of a formulation, it is necessary to address the effects of accidental freezing of product during shipping and inadvertent freezing of solution formulations intended to be stored at refrigerated temperatures. In addition, since some excipients can precipitate upon freezing, it is important to reexamine the freeze/thaw issue once a formulation has been chosen.

D. Organic Solvent Compatibility

To gain a preliminary understanding of the compatibility/stability of the protein with organic solvents it is helpful to conduct a small compatibility study. The results can help in the selection of analytical methods, establishment of preparative chromatographic techniques, and in setting up processing conditions for incorporation into drug delivery matrices. Table 5 shows data for a recombinant glycoprotein at a concentration of 0.5 mg/mL in a pH 6.0 phos-

Table 5 Organic Solvent Compatibility Data for a Recombinant
Glycoprotein

| Solvent composition | Content (%)[a] | |
	Monomer	Dimer/Aggregate[b]
Phosphate buffer	100	0
Methanol 50%	100	0
Ethanol		
15%	100	0
30%	98	2
40%	96	4
50%	95	5
n-Propanol		
35%	100	0
50%	69	31
Isopropanol		
35%	100	0
50%	97	3

[a] Determined by size exclusion chromatography.
[b] Dimer was present only in the 50% n-propanol sample.

phate buffer that was exposed to a variety of solvents at several concentrations
for 4 hours at 4°C storage conditions. These data were generated using SEC
coupled to an ultraviolet detector to monitor changes in the aggregation state
of the molecule. All solvents were moderately compatible up to approximately
35% organic solvent; from a stability standpoint, the product possessed sufficient compatibility to examine several of these solvents for use as eluents from
a chromatographic column.

E. Preservative Screening with DSC

As previously stated, determination of the melting temperature for a particular
protein in the presence and absence of a potential formulation component can
be helpful during the preformulation process. The effects of various ingredients on the thermal denaturation of the protein may then be used as one method
for quickly assessing excipient compatibility and the likelihood of achieving
long-term protein stability in a particular formulation. Table 6 summarizes
the characterization of experimental values of the melting temperature of a

Table 6 Effect of Benzyl Alcohol Concentration on
the Melting Temperature of a Recombinant Protein

Benzyl alcohol concentration (%)	$T_m(°C)$ at two protein concentrations	
	0.4 mg/mL	16 mg/mL
0.0	60.0	59.7
0.5	57.1	56.6
1.0	53.9	51.0

recombinant glycoprotein in the presence and absence of benzyl alcohol, to assess the potential of achieving a long-term, stable formulation containing benzyl alcohol as a preservative.

In the absence of benzyl alcohol, the protein undergoes a single endothermic transition at 60°C. Increasing amounts of benzyl alcohol produced a corresponding decrease in T_m for both protein concentrations. Protein instabilities, such as changes in tertiary structure, aggregation, decreased solubility, and reduction in melting temperature, have been reported in formulations containing benzyl alcohol (23–25). The mechanism of destabilization in this particular case is unclear when only the DSC data are considered; however, circular dichroism experiments have shown benzyl alcohol to produce tertiary structure changes through direct interaction with some proteins (23,24).

F. Identification of Key Degradation Mechanisms

As discussed earlier, the identification of the degradation mechanisms a protein undergoes allows rational design decisions during formulation development. At this stage of early development, it is sufficient to identify the mechanisms of degradation, leaving the assignment of the specific amino acids involved in a reaction until later in development if the degradation is found to persist in the final formulation. Either of two approaches can be used in identifying reaction mechanisms: identification of the degradation product or identification of a by-product of the reaction. To rationally design a formulation for the monoclonal antibody shown in the first case study, it was important to positively identify the cause of the shifts in the IEF pattern. When these experiments were performed, neither a tryptic map for the antibody nor a deamidation-specific analytical method had been established. Instead, a surrogate assay for deamidation, which was thought to be the most likely degradation mechanism, was performed. Specifically, the amount of ammonia liberated from the antibody sample was studied at pH 9.0 for several weeks at room temperature; the results are plotted in Fig. 8. Increasing concentrations of ammonia were detected over this time period, which correlated with changes in the IEF pattern of the protein from high to low pI values, both of which are indicative of deamidation. At this early stage of development, such surrogate assays require little development time and serve to confirm the mechanism of degradation. In the final formulation, if this degradation were to persist, assignment of the specific asparagine or glutamine amino acid(s) undergoing deamidation would have to be made by more specific analytical methods such as tryptic peptide sequencing (5). From these data, deamidation was established as a primary degradation mechanism for this protein, and formulation studies examined parameters such as pH, salt content, and specific buffer species, all of which have been shown to affect the rate of deamidation reactions (15).

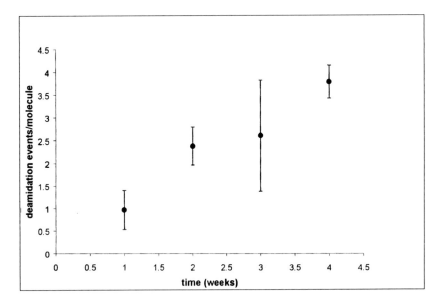

Fig. 8 Enzymatic determination of ammonia released during storage of a monoclonal antibody at pH 9.0 and 25°C. Quantitation performed using Sigma Diagnostics Ammonia Kit.

G. Restarting the Preformulation Process

It is important that the formulator be willing to restart the preformulation process at any time during product development. During initial formulation development, a lyophilized dosage form was developed that possessed acceptable stability for use in early clinical trials. Later in development, the formulation was changed for delivery purposes, which required that moisture be introduced into the formulation. This change in the formulation required that the preformulation process be restarted to examine the effect of moisture on the solid state stability of the protein. Initially, it was thought that because the protein had acceptable stability in both lyophilized and solution formulations, the most significant problem likely to be encountered was changes in the bulk properties of the powder that could affect delivery from the gelatin capsule.

The first set of preformulation experiments performed examined the effect of different temperature/moisture conditions on the dose delivered from the gelatin capsule. As expected, the level of moisture in the powder formulation dramatically affected reproducibility of the dose delivered from the capsule. However, during these experiments, it was also observed that exposure of the powder to moisture resulted in dramatic changes in the aggregation state of the protein when analyzed by size exclusion chromatography and SDS-PAGE. Exposure of the dry lyophilized cake to moisture, specifically 25°C at 60% relative humidity, caused a fourfold increase in soluble protein dimer by SEC. Further study of this aggregation phenomenon led to the conclusion that the aggregation process was due to disulfide bond exchange, which was greatly accelerated in the lyophilized powder formulation with the addition of moisture, as reported earlier for other proteins in the solid state (26). By controlling the level and duration of moisture exposure, this aggregation process and the changes in the bulk properties of the powder were minimized. This detection of protein aggregates led to a redevelopment of the SEC technique using a more sensitive detection method and incorporation of a chromatography column capable of higher resolution between dimer and monomer species.

Retrospective analysis of earlier formulations with the modified SEC method revealed that the disulfide scrambling and resulting dimer formation were not unique to the powder formulation upon exposure to moisture. When the modified SEC technique was used, low levels of dimer were detected in samples made and stored under vacuum with only 3% residual moisture after lyophilization. These low levels of aggregate ($< 0.5\%$) were also masked by coelution of an excipient in the lyophilized formulation. Examination of solution formulations also revealed about 0.5% of the protein aggregate. This process of restarting preformulation experiments during later development resulted in a better understanding not only of the handling and processing conditions for producing the powder formulation, but also of the mechanism underlying the instability, which was also applicable to other related formulations. This led to further refinement of the analytical methodology for monitoring protein stability.

VI. CONCLUSIONS

Throughout this chapter we have attempted to give the formulation scientist a step-by-step approach for performing a preformulation study on a protein drug candidate. Preformulation should be an iterative process that starts early in development and continues through the development life of a molecule. Preformulation studies are essential for successful product development, and a formulator should be willing to restart preformulation efforts when business and marketing decisions dictate a change or modification of a formulation.

ACKNOWLEDGMENTS

The authors gratefully acknowledge the efforts of the protein formulation group in generating much of the data that have made this chapter possible. Specifically, we thank Gerry Bell, Tobin Cammett, Neil Gesuero, Christine Hall, Alison Mares, Paul McGoff, David Scher, Dr. Bret Shirley, and Glenn Snow.

REFERENCES

1. Lucian C. Chen, Traditional techniques enhance advances in biotherapeutic formulations. Gen. Eng. News, July 23, 1994, p. 170.
2. J. M. Piet Van den Oetelaar, et al., Stability assessment of peptide and protein drugs. J. Controlled Release 21:11–21, 1992.
3. Brigitte M. Eckhardt, James Q. Oeswein, and Thomas A. Bewley, Effect of freezing on aggregation of human growth hormone. Pharm. Res. 8(11):1360–1364, 1991.
4. M. V. Paranandi, A. W. Guzzetta, W. S. Hancock, and D. W. Aswad, Deamidation and isoasparate formation during in vitro aging of recombinant tissue plasminogen activator. J. Biol. Chem. 269:243–253, 1994.
5. Dana W. Aswad and Andrew W. Guzzetta, Methods for analysis of deamidation

and isoaspartate formation in peptides and proteins. In Deamidation and Isoasparate Formation in Peptides and Proteins, Dana W. Aswad ed. CRC Series in Analytical Biotechnology, CRC Press, Boca Raton, FL, 1995.

6. Antonio Sillero and Joao Meireles Ribeiro, Isoelectric points of proteins: Theoretical determination. Anal. Biochem. 179:319–325, 1989.

7. Andrew D. Robertson and Kenneth P. Murphy, Protein structure and the energetics of protein stability. Chem. Rev. 97:1251–1267, 1997.

8. Kunihiko Gekko and Serge N. Timasheff, Thermodynamic and kinetic examination of protein stabilization by glycerol. Biochemistry 20:4677–4686, 1981.

9. Wayne R. Gombotz, Susan C. Pankey, Lisa S. Bouchard, Duke H. Phan, and Alan P. MacKenzie, Stability, characterization, formulation, and delivery system development for transforming growth factor-beta₁. In: Formulation, Characterization, and Stability of Protein Drugs, Rodney Pearlman and Y. John Wang, eds. Plenum Press, New York, 1996, pp. 219–245.

10. James C. Lee and Serge N. Timasheff, The stabilization of proteins by sucrose. J. Biol. Chem. 256(14):7193–7201, 1981.

11. Lucy L.-Y. Lee and James C. Lee, Thermal stability of proteins in the presence of poly(ethylene glycols). Biochemistry 26:7813–7819, 1987.

12. Mark C. Manning, Kamlesh Patel, and Ronald T. Borchardt, Stability of protein pharmaceuticals. Pharm. Res. 11(6):903–918, 1989.

13. R. Pearlman and T. Nguyen, Pharmaceutics of protein drugs. J. Pharm. Pharmacol. 44:178–185, 1992.

14. Babur Z. Chowdhry and Steven C. Cole, Differential scanning calorimetry: Applications in biotechnology. Trends Biotechnol. 7:11–18, 1989.

15. Todd V. Brennan and Steven Clarke, Deamidation and isoasparate formation in model synthetic peptides: The effect of sequence and solution environment. In: Deamidation and Isoasparate Formation in Peptides and Proteins, Dana W. Aswad, ed, CRC Series in Analytical Biotechnology, CRC Press, Boca Raton, FL, 1995.

16. D. E. Staunton, S.D. Marlin, C. Stratowa, M.L. Dustin, and T.A. Springer, Primary structure of ICAM-1 demonstrates interaction between members of the immunoglobulin and integrin supergene families. Cell 52:925–933, 1988.

17. A. F. Henson, J. R. Mitchell, and P. R. Mussellwhite, The surface coagulation of proteins during shaking. J. Colloid Interface Sci. 32:162–165, 1970.

18. M. G. Mulkerrin and R. Wetzel, pH dependence of the reversible and irreversible thermal denaturation of gamma interferons. Biochemistry 28:6556–6561, 1989.

19. Murray P. Deutscher, Maintaining Protein Stability. Methods Enzymol. 182:83–89, 1990.

20. Sandeep Nema and Kenneth Avis, Freeze–thaw studies of a model protein, lactate dehydrogenase, in the presence of cryoprotectants. J. Parenter. Sci. Technol. 47(2):76–83, 1993.

21. Peter L. Schwartz and Margaret Batt, The Aggregation of [¹²⁵I] human growth hormone in response to freezing and thawing. Endocrinology 92(6):1795–1798, 1973.

22. Byeong S. Chang, Brent S. Kendrick, and John F. Carpenter, Surface-induced

denaturation of proteins during freezing and its inhibition by surfactants. J. Pharm. Sci. 85(12):1325–1330, 1996.

23. Jonas Fransson, Dan Hallen, and Ebba Florin-Robertsson, Solvent effects on the solubility and physical stability of human insulin-like growth factor I. Pharm. Res. 14(5):606–612, 1997.

24. Xanthe M. Lam, Thomas W. Patapoff, and Tue H. Nguyen, The effect of benzyl alcohol on recombinant human interferon-γ. Pharm. Res. 14(6):725–729, 1997.

25. Wayne R. Gombotz and Richard L. Remmele, DSC as a protein formulation tool: Comparisons to real time stability studies. Revolutions in Stability Testing: Condition, Analytical Methods, and Regulatory Perspectives—Part II. Biologicals, American Association of Pharmaceutical Scientists, Annual Meeting and Exposition, Boston 1997.

26. Robert W. Liu, Robert Langer, and Alexander M. Klibanov, Moisture-induced aggregation of lyophilized proteins in the solid state. Biotechnol. Bioeng. 37: 177–184, 1991.

5

Solution Formulation of Proteins/Peptides

Paul McGoff and David S. Scher
Alkermes, Inc.
Cambridge, Massachusetts

I. INTRODUCTION

The simplest and most economical way to formulate a protein is to develop a solution formulation. Lack of adequate protein stability in solution may ultimately require development of an alternative, more complex formulation such as a lyophilisate. A well-designed formulation study will allow the formulation scientist to determine whether a solution formulation will be acceptable for a given protein.

The first requirement for a formulation study is the availability of significant quantities of purified protein and/or peptide. Along with this protein or peptide should come a vast amount of practical knowledge obtained during various stages of process development. This knowledge will not only be anecdotal in nature but will also provide some specific physicochemical characteristics of the molecule. Purification in-process analytical tests should be available to assess basic parameters such as purity, concentration, and some measure of activity. Experiences during early stages of process development can contribute to this initial knowledge base. For example, problems such as poor solubility or aggregation that may occur during fermentation, harvesting, and purification of the molecule can result in a reduced recovery yield. Knowledge of the conditions and circumstances under which these problems occurred can aid in focusing and prioritizing the initial formulation studies. For example, a particular pH or mix of buffer components used during processing may reduce solubility or induce aggregation. Even physical processing steps like diafiltration can expose the protein to shear and denaturing liquid–solid or liquid–air interfaces, which can ultimately cause protein aggregation as well. Alternatively, investigating the effect of freezing might be advisable, since freezing is often used as a convenient process-hold step. Feedback from these experiences can be used to design formulation experiments to gain insight on how to avoid, minimize, or explain process problems. Conversely, the results from these formulation experiments can aid in refining the purification process as well as in guiding future formulation development to address specific potential stability problems.

The second prerequisite for starting protein solution formulation studies is availability of analytical test procedures for characterizing the physicochemical properties of the protein. Methodologies for the characterization, beyond the basic ones used to assess purity during stages of purification, are necessary. As part of preformulation studies, specific stability indicating analytical methods will need to be developed. These methods must be capable of detecting changes in protein samples that have been altered by physical or chemical stress. Typical properties monitored include changes in charge [studied by means of isoelectric focusing (IEF), native–polyacrylamide gel electrophoresis (native-PAGE), and ion exchange chromatography], conformation [size exclusion chromatography (SEC), circular dichroism, fluorescence spectroscopy, native-PAGE, and capillary zone electrophoresis], size [SEC, sodium dodecyl sulfate–polyacrylamide gel electrophoresis (SDS-PAGE), laser light scattering, matrix-assisted laser desorption ionization time-of-flight (MALDI TOF) mass spectrometry, native-PAGE, and capillary electrophoresis], hydrophobicity (hydrophobic interaction chromatography, reversed phase chromatography, and micellar electrokinetic chromatography), and biological activity (cell-based assay, enzymatic assay, and enzyme-linked immunosorbant assays).

Additionally, knowledge of a protein's unique structural and functional characteristics can often indicate which analytical tests will be most useful in assessing stability. For example, when one is probing for changes that effect hydrophobicity, such as denaturation or methionine oxidation, hydrophobic interaction chromatography is generally more useful for large proteins (immunoglobins), whereas reversed phase techniques are more appropriate for small proteins (MW < 40 kD) and peptides. Many glycosylated proteins contain negatively charged sialic acid residues, which create surface charge heterogeneity and the potential for complicated IEF gel patterns and poorly resolved native-PAGE gels. Such complexity might mask charge changes that one might expect due to deamidation, for example. Extensive glycosylation can also result in size heterogeneity that may contribute to zone broadening in SDS-PAGE or SEC. Calculations based on amino acid composition or sequence can be used to estimate charge as a function of pH (including isoelectric point) for a particular protein (1). This is useful in choosing electrophoretic or ion exchange conditions and may predict solubility behavior, since solubility can be strongly influenced by charge. Coupling specific structural knowledge of a protein with general knowledge of protein degradation will help in choosing the most appropriate analytical methods for a particular protein. In addition, it will give one an idea of what results to expect when a particular analytical method is applied.

Consideration of all the aforementioned prerequisites will allow for some prioritization and flexibility in designing solution formulation studies. In the end, formulation studies should address solubility as a function of pH and salt and the influence of increased temperature, solution pH, buffer ion, salt, protein concentration, and other excipients and preservatives (when necessary) on stability. Other studies to include are photostability, cavitation/ shaking, and freeze–thaw cycling. Finally, material compatibility studies should be performed with any storage containers or medical device the molecule/formulation may contact.

II. SOLUTION FORMULATION

A. Solubility Studies

Typically, the relevant pH range over which solubility needs to be determined, for protein formulation studies, is 4 to 9. Although processing may expose the protein or peptide to pH values varying from 2.5 to 11, extremes in pH are avoided in formulations, since many degradative processes become more prevalent below pH 5 (acid hydrolysis) and above pH 7 (deamidation). To assess the minimum solubility required to formulate and deliver a protein drug, it is useful to consider the maximum dosing limit. A survey of the PDR (*Physician's Desk Reference*) reveals that for direct injection intravenous (IV) formulations, the upper limits on the volume and dose for currently marketed protein therapeutics, are 10–20 mL and 10 μg–100 mg of protein, respectively. Therefore, this means that one may need to achieve a minimum solubility in the range of 0.1–5 mg/mL. If the solubility is less than 0.1 mg/mL in this pH 4–9 range, additional excipients or sodium chloride may need to be examined as a means of increasing solubility.

If lyophilized, excipient-free protein is not readily available or obtainable owing to specific protein chemistry considerations, an alternative (and more laborious) way to perform solubility studies is to diafilter a protein stock solution into the desired buffer and then concentrate it as high as is practically possible. A variety of small-volume, centrifuge-driven, pressure-driven, and vacuum-driven devices are commercially available for this purpose. The concentrated stock solution can then be diluted with the appropriate series of buffers to be used in the solubility studies. An example of a solubility study protocol is as follows:

> A ready supply of either lyophilized, excipient-free protein or diafiltration equipment is needed.
> Start with protein in dry form and incrementally add weighed amounts to a small volume (typically ≤ 1 mL) of test buffer.
> When the solubility limit is exceeded (i.e., when precipitation occurs), the undissolved protein pellet is centrifuged and the supernatant is analyzed for concentration.
> Limitations on the amount of protein to add are discretionary as long as acceptable solubility is achieved.

Typically, if protein solubility exceeds 50 mg/mL at the pH of interest, no further solubility studies are necessary. The solubility desired should be minimally two fold and maximally 10-fold higher than the maximum conceivable dose.

If protein solubility is limited at relevant formulation pH values, additional measures will need to be taken to increase solubility. Some of the ways in which solubility can be increased include increasing/decreasing sodium chloride, adding other salts, varying the buffer species at a given pH, or including glycerol, lipids, polymers, cyclodextrins, or surfactants. Generally, one or the other of these approaches has successfully been applied to increasing solubility in protein solution formulations (2,3).

B. Isotonicity Considerations

Since isotonicity alone may have a tremendous influence on solubility, the basic approach should be to adjust the sodium chloride concentration before other excipients are added to the formulation. From a physicochemical standpoint, it is best to have as simple a formulation as possible, with fewer opportunities for potential complications due to undesired excipient–protein interactions. Furthermore, when problems do arise, sorting out what is happening is much easier. When considering salt concentration, our recommendation is to formulate at or near isotonic conditions. While hypotonicity is not typically a problem for injectables, nasals, and topical products, hypertonic solutions may cause undesirable local tissue irritation or a burning sensation upon administration.

C. Solution Stability Studies

Generally, a formulation pH stability study should be limited to the physiologically relevant pH range of 4–9. Preformulation studies will typically address conditions and degradation pathways outside this pH range. The choice of buffer species should be physiologically compatible and listed as GRAS (generally regarded as safe) by the U.S. Food and Drug Administration. A useful source of excipients used in formulations is the FDA guide to inactive ingredients (4). Often at a given target pH, where multiple buffer choices are available, there will be differences between specific buffer ions with respect to effect on certain degradative pathways or rates of degradation.

D. Specific Buffer Ion Effects

To illustrate the importance of considering the specific ion effects of the buffer chosen, we discuss an example for a nine amino acid peptide (RMP-7). One clinical formulation for RMP-7 drug product consisted of a pH 4.0 unbuffered solution in normal saline (0.9% sodium chloride, pH adjusted with acid/base). This product was reformulated into a buffered system at the same pH to minimize pH drift due to leaching of hydroxyl ions from the surface of the glass vials of the drug product .The two choices for buffer species were citrate (pK 2.5, 4.5, 6.0) and acetate (pK 4.5), both of which possess good buffering capacity near the target pH of 4.0. The study consisted of formulating RMP-7 drug substance in a series of citrate buffers ranging in pH from 2.5 to 6.0 and in acetate buffers at pH 4.0,4.5, and 5.0. Solutions were stored for up to 1 month at 60°C in a well-controlled temperature stability cabinet. Potency and purity of RMP-7 solution formulations were assessed as a function of time using a validated reversed-phase high performance liquid chromatography (HPLC) assay. Figure 1 graphically illustrates the effect of the two buffer species on the stability of RMP-7, as measured by drug purity as a function of solution pH verses the unbuffered formulation. In this case, the use of citrate ion caused more rapid peptide degradation than the acetate formulations over the pH range of 4.0–5.0. The acetate formulations were equivalent to the unbuffered formulation at pH 4.0.

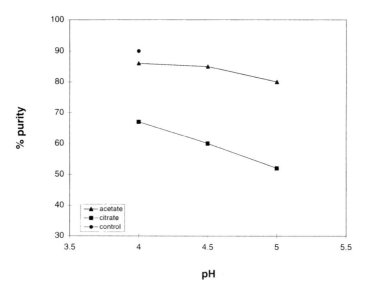

Fig. 1   Degradation of the peptide RMP-7 as a function of pH in two types of equivalent molarity buffer. The control solution was unbuffered with pH adjusted to 4 by addition of NaOH and HCl. All solutions were adjusted with NaCl to isotonicity and stored for one month at 60°C in molded glass vials. The difference in degradation rate demonstrates the buffer ion effect on RMP-7 stability.

This was an important finding in the formulation development of RMP-7 and illustrates the importance of choosing the appropriate buffer species for a formulation. This choice will always be dependent on the characteristics of the individual drug substance and cannot easily be predicted in advance nor generalized to other drugs. Once the likely pH of the final product formulation has been determined (from earlier preformulation studies), testing more than one buffer species at multiple concentrations at the chosen pH is advisable.

To measure the buffer ion effect on a degradative pathway, the buffer ion concentration in a formulation is varied and the tonicity is kept constant by addition of sodium chloride. A plot of the observed degradative rate constant versus buffer concentration will have a slope proportional to the buffer ion activity. Extremes in temperature and pH will accelerate degradation. Study design will require frequent sampling (i.e,. hourly). Once the pH range has been narrowed to where major degradative processes are slowed, if degradation is still unacceptable (projected shelf life at 4°C <2 years), the effects of adding other potentially stabilizing excipients should be studied. However, if projected degradative rate is acceptable within this narrowed range, formulation choice becomes more discretionary and can be based upon experience and other dosing or delivery considerations. With proteins, one should be cautious about projecting 4°C stability from accelerated temperature studies. At elevated temperatures other pathways/mechanisms can become more prevalent, often confusing analytical results and resulting in erroneous shelf life predictions. Some degradative processes have nonlinear plots of rate as a function temperature or may have plateaus, complicating projection [5–9].

E. Excipients

The inclusion of stabilizing excipients leads to the most interesting phase of formulation development studies. Here exists a vast sea of folklore and anecdotal information alongside excellent case history literature (3). Personal preferences often dictate excipient choices based on past experiences with different proteins (not always with positive results). Every protein has unique physicochemical characteristics and the potential to behave differently stability-wise in the presence of a particular excipient. For example, the common usage of surfactants to stabilize small recombinant proteins does not automatically warrant this inclusion in monoclonal antibodies formulations. Excipients, if necessary, should be chosen for their known functionality in slowing or arresting specific degradative pathways. An example of specific excipient choice based on known functionality is the use of hydroxypropyl-β-cyclodextrin in stabilizing interleukin (IL-2) aggregation (2,10).

F. Temperature Stability

Normally, proteins in solution are formulated for storage at 4°C with a minimum target shelf life of 2 years. As stated in the International Conference Harmonization (ICH) stability guideline (11):

> Since most biotechnological/biological products need precisely defined storage temperatures, the storage conditions for the real-time/real temperature stability studies may be confined to the proposed storage temperature. However, it is strongly suggested that studies are conducted on the drug substance and drug product under accelerated and stressed conditions. Studies under stress conditions may be useful in determining whether accidental exposure to conditions other than those proposed (e.g. during transportation) are deleterious to the product. Conditions should be carefully selected on a case-by-case basis.

Most solutions of proteins are not thermally stable above room temperature for prolonged periods of time; therefore accelerated studies are typically conducted at 25, 30, and/or 40°C, and generally no higher. However, in the case of peptides, accelerated studies at higher temperatures are often permissible because their inherent temperature stability is greater. Conducting a temperature rate study of degradation may allow not only a projection of stability at 4°C, but also an estimation of the activation energy of a particular process (12).

G. Protein Concentration

Another important formulation parameter, that needs to be studied is the effect of protein concentration on stability. Of course early on in development, availability of sufficient amounts of protein active may limit the extent to which high concentration studies can be conducted. However, such studies should be performed because rates of degradation typically vary as a function of protein concentration. Based on overall clinical experience with protein drugs and their maximum inherent solubilities, we recommend starting in the 1–10 mg/ mL protein concentration range. This assumes that the final drug product will be administered by IV injection in a volume of 1–10 mL.

H. Photostability

Another ICH guideline states: "The intrinsic photostability of new drug substances and drug products should be evaluated to demonstrate that, as appropriate, light exposure does not result in unacceptable change. Normally, photostability testing is carried out on a single batch of material" (13). This particular test will be most relevant for products formulated as solutions where there is potential for storage under lighted conditions. Since most biologicals are stored refrigerated (the light goes out when you close the door), the assessment of photodegradation is a minor consideration, but nonetheless a regulatory requirement. After controlled light exposure of RMP-7 drug product, the nine amino acid peptide discussed earlier, an increase in one impurity was observed, but the product still met acceptance criteria of ≤1% impurity. This is in contrast to RMP-7 drug substance, which exhibited no photodegradation when exposed to the same conditions.

I. Preservatives

1. General Considerations

If the protein formulation is intended for multidose use, a preservative will need to be included to prevent microbial growth. However, preservatives are toxic, and there are maximum limits described for injectables (14). Table 1, assembled from a search of the PDR, lists preservatives used in some approved protein and peptide pharmaceuticals. Selection of an appropriate preservative is not trivial and requires the consideration of many factors, which complicates formulation. Focus initially will be to choose several potential preservative candidates based on current use and route of drug delivery, ideal solution pH of formulation, and compatibility with other formulation components. Once a selection has been made, accelerated and real–time stability testing of the protein in the presence of these selected preservative candidates should be performed. Chapter 4 on preformulation contains a discussion on the use of differential scanning calorimetry to screen preservatives with respect to protein interactions. In addition, an article on the effect of benzyl alcohol on the stability of γ-interferon is available (15). These authors do not identify the mechanism of interaction but do show that by picking the correct buffer salt and minimizing the amount of preservative, a stable formulation can be developed.

2. Timing for Inclusion in a Formulation

Performing extensive stability testing of the protein without a preservative will provide one with an idea of the buffer system and pH on which to focus stability testing. If the objective of the initial phase I clinical trial is proof of concept, the simplest approach is formulation without a preservative. However, this approach will mean repeating the phase I trial with a second preserved formulation, and additional stability studies will be required. If it is known that the final drug product will be used in a multidose form, a preserved formulation must be developed at some point. The timing of the decision to develop and test the preserved formulation will depend heavily on a cost–risk benefit assessment.

A decision to develop a preserved formulation will require developing methods to assay both protein and preservative. Processing equipment and container compatibility with the preservative, as with the protein, will need to be addressed. Finally, preservative efficacy in the protein formulation will need to be performed on the preservative candidates.

Quick elimination of a particular candidate formulation due to stability problems will reduce the amount of samples that need to be tested by preservative efficacy.

3. Choosing a Preservative

There are a number of prerequisites for selecting a potential preservative candidate. Is it generally regarded as safe and pharmaceutically acceptable

Table 1 Currently Used Preservatives for Proteins and Peptides from the PDR

Product, manufacturer, and drug	Route of administration[a]	Container	pH[b]	Buffer system[b]	Preservative concentration
Proteins					
Alferon N, Interferon Sciences, Inc.; interferon alfa-n3	IL	1 mL vial	7.4	Sodium and potassium phosphate	Phenol: 3.3 mg/mL
Epogen, Amgen, recombinant erythropoietin (epoetin alfa)	IV or SC	2 mL vial	6.1	Sodium citrate	Benzyl alcohol: 1%
Intron A, Schering; recombinant interferon alfa-2b	IM, SC, or IL	0.5, 1.0, and 3.2 mL vials		Sodium phosphate	1.5 mg/mL m-cresol, 0.1 mg/mL EDTA
Nutropin AQ, Genentech; somatropin	SC	2 mL vial	6.0	Sodium citrate	2.5 mg/mL phenol
Procrit, Ortho Biotech; recombinant erythropoietin (epoetin alfa)	IV or SC	2 mL vial	6.1	Sodium citrate	Benzyl alcohol: 1%
Roferon-A, Roche; recombinant interferon alfa-2a	SC or IM	1, 0.9, and 3mL vials	NA	NA	Phenol: 3.3 and 3.0 mg/mL

Peptides

Calcimar, Rhone Poulenc; calcitonin	SC or IM	2 mL vial	NA	Sodium acetate	Phenol: 5 mg/mL
Desmopressin, Ferring; desmopressin acetate	IN and IV	2.5 and 10 mL vial	4	No additional	Chlorbutanol: 5 mg/mL
DDAVP injection, Rhone Poulenc; desmopressin acetate	IV	10 mL vial	4	No additional	Chlorbutanol: 5 mg/mL
DDAVP nasal spray, Rhone Poulenc; desmopressin acetate	IN	2.5 mL vial, 5 mL bottle	NA	Citric acid, sodium phosphate	Benzalkonium chloride; 0.1 mg/mL
Iletin II pork, Lilly; pork insulin	SC	20 mL vial	NA	No additional	m-cresol: 0.25% (w/v)
Miacalcin, Novartis Pharmaceuticals; calcitonin	SC or IM	2 mL vial	NA	Sodium acetate	Phenol: 2.25 mg/mL
Miacalcin, Novartis Pharmaceuticals; calcitonin	IN	2 mL vial	NA	No additional	Benzalkonium chloride: 0.1 mg/mL
Pitocin, Parke Davis; oxytocin	IV	10 mL vial	NA	Acetic acid	Chlorbutanol: 0.5%
Sandostatin, Novartis Pharmaceuticals; octreotide acetate	Deep SC or IV	5 mL vial	4.2	Lactic acid	Phenol: 5 mg/mL
Synarel, Searle; nafarelin acetate	IN	10 mL bottle	NA	Acetic acid	Benzalkonium chloride

[a] IN, intranasal; IV, intravenous; IL, intralesional; SC, subcutaneous.
[b] NA, not available.

(GRAS-listed)? The intended route of administration is also an important consideration. A preservative appropriate for IV administration may not be appropriate for nasal administration. Some preservatives that are acceptable for IV administration may have a strong odor or taste (phenolics); others may dry nasal mucosa or be cilia static or toxic (16). Sensitivity of target patient population also may be an important consideration. For example, asthmatics have been shown to be more sensitive to certain types of preservative. Benzalkonium chloride and EDTA from nebulized solutions have both been reported to induce dose-related bronchoconstriction in asthmatics (17).

Formulation pH is another important consideration in choosing a preservative. Preservatives are most effective against microbes within specific pH ranges. When formulated outside these ranges, many preservatives change ionization state and lose efficacy. The candidate should have a wide spectrum of activity, preventing the growth of fungi, as well as exhibiting cidal activity against various types of Gram-positive and -negative bacteria. Preservatives can often be used in combination to achieve synergistic effects (18,19).

Compatibility with other formulation components is another important consideration. Certain preservatives (e.g., mercurial preservatives are incompatible with metals; others (e.g.) phenolics, quaternary ammonium, paraben preservatives), are incompatible with surfactants. Choice of preservative concentration to use can initially be based upon current use in the industry (PDR) and specific literature references for preservative safety and efficacy against various types of microbe (19). Eventually, preservative efficacy studies performed with the protein will determine the optimum preservative concentration.

4. Analytical Methods

Many preservatives have characteristics that interfere with analytical methods that work extremely well for the protein in their absence. For example, many preservatives exhibit high ultraviolet adsorption, making it necessary to have analytical techniques capable of separating the protein active from the preservative. Often the preservative needs to be removed before cell-based activity assays are performed, since the preservatives are toxic to the cells. Ideally, it would be nice to have quantitative separation methods capable of resolving the preservative(s) from the protein with no interfering interactions from any of the components. We have seen that paraben preservatives interact with hydrophilic (SEC) media, complicating results. In contrast, a reversed-phase method that worked well to separate and quantitate parabens irreversibly adsorbed the protein and thus modified column performance. Developing analytical techniques to get around these problems can be a challenge.

5. Compatibility

In addition, container considerations may change when a preservative is added to a protein formulation. Some preservatives are sensitive to light and/or air and may require storage in brown glass containers and/or airtight or purged containers. Phenolics, benzyl alcohol, chlorobutanol, parabens, and mercurial preservatives are examples.

Container–preservative compatibility can be a problem. Parabens and benzalkonium chloride adsorb to and chlorobutanol diffuses through plastic containers and delivery devices, reducing preservative strength (20,21). Furthermore, rubber stoppers adsorb many preservatives. Additionally adsorption/interaction with processing equipment that contacts preservative, such as vessels, tubing, and filters, can be potentially problematic. Adsorption studies should be performed to address these potential pitfalls.

6. Preservative Efficacy

Preservative efficacy studies are performed on preserved formulations as described under the following general test chapter of the U.S. Pharmacopeia: *Antimicrobial Preservatives–Effectiveness* •51® (USP 23, p. 1681) and are useful in identifying the most appropriate preservative for a particular protein at the optimum preservative concentration to maintain a growth-free solution. The assay involves challenging the formulation with bacteria of three types (*Staphylococcus aureus*, *Escherichia coli*, and *Pseudomonas aeruginosa*) and fungi of two types (*Candida albicans* and *Aspergillus niger*). The effective concentration range of the preservative is determined by adding various levels to the formulation and performing USP ⟨51⟩. For example, if the intended formulated preservative concentration of phenol is 0.3 % w/v, the effectiveness of this concentration is tested at this level and several lower levels (e.g., 0.2, 0.15, 0.1 % w/v). The lower concentration levels are tested to ensure a margin of safety, since preservatives can become less effective owing to either degradation or adsorption over time.

J. Cavitation/Shaking

Proteins in solution denature at air liquid interfaces. Under significant stress— for example, during filtration or physical shaking—a portion of the soluble protein will often denature and irreversibly aggregate. A typical result of this denaturation process is a turbid, cloudy solution or a solution containing insoluble protein particles. Therefore, it is a good idea to test the ruggedness of a formulation to withstand shaking or agitation that that might occur during handling. To test a formulation's ruggedness, ultimately the final dosage form (in its final container) should be shaken or cavitated. A mechanical shaker is used to cavitate the protein solution for days and up to a week at room temperature. If the solution becomes cloudy or SEC reveals soluble aggregates, adding volume to fill container often reduces the problem. Another approach would be to supplement the formulation by adding glycerol, polymer, or a surfactant, but this ultimately depends on the sensitivity of the protein. If the solution does become cloudy, consider adding the instruction "DO NOT SHAKE" to the container label.

K. Freezing Studies

Freezing is another handling consideration. At various stages in processing, it may be necessary to freeze bulk protein solutions. Although the storage condition of final drug product is generally 2–8°C, there is always the possibility of inadvertent exposure to freezing temperatures during shipping and storage. Freezing might occur during air shipments of protein solutions, since packages may end up in uncontrolled-temperature cargo bays or in refrigerated storage units having colder areas that are below freezing. Thus it is important to do freeze–thaw cycling studies to determine the ruggedness of the protein solution. In the processing of proteins, the availability of containers of many types and sizes makes it a challenge to the design a relevant scalable freeze–thaw simulation at the small scale. Variations in the process of freezing can radically affect the rate of freezing and ultimately the integrity of the protein. The temperature of the freezer ($-20°C$ or $-80°C$), the type of freezer, the method of freezing (contact of shelf, liquid, or air), the location within the freezer, the volume of solution, and the container geometry all affect the way in which the solution freezes. Details of how these parameters affect protein integrity should be carefully considered in freeze–thaw cycling experimental design. One approach is the use of minitanks to assess the potential damage caused by repeated freeze–thaw cycles in a scaled-down version of process tanks. Studies of these types have been used to support the freezing and subsequent thawing of bulk drug substances stored in large-scale tanks. Protein solutions are analyzed both visually and by (SEC) for aggregation.

For final drug product, such studies are somewhat easier, because the container is a limiting constraint. Freeze–thaw studies using the final drug product are considered relevant because product may encounter hostile environmental conditions en route to its destination or accidentally in the hands of health care workers (doctors, nurses, pharmacists).

L. Materials Adsorption

Materials adsorption is mainly a consideration for low concentration protein/peptide formulations, generally at concentrations less than 1 mg/mL. Increasing surface area, contact time, and temperature all can enhance adsorption. Adsorption has the potential to occur during processing and during storage in final drug product container. Materials that contact the protein solution during processing stages and in the final container will determine what compatibility testing needs to be done. Table 2 shows what types of material may need to be tested and at the stage during processing at which contact occurs.

If the active drug is formulated at a low concentration and adsorbs to the container of choice, there are several options. One option is to fill the container to maximum volume. Filling a container to near capacity can reduce surface area contact between protein/peptide solution and the container, thus minimizing potential adsorption. The second and more difficult option is to change container material type. As a last resort, one may consider the addition of another excipient (e.g., human serum albumin, polymers, or surfactants) to the formulation, to inhibit absorption.

M. Placebo Considerations

In developing placebos for protein solution formulations, one must consider that the properties of the placebo solution will differ slightly from those of the active substance. There will be a difference in the solution viscosity, and a shaken protein solution will exhibit a certain degree of foaming, which may allow a clever clinician the opportunity to unblind a trial. As far as stability is concerned there should be no difference between the placebo and active solutions. To demonstrate this, it is prudent to manufacture and monitor the stability of the placebo product prior to initiating clinical trials. This will en-

Table 2 Material Contact at Various Processing Stages

Surface material	Drug substances	Formulated bulk	Final drug product
Plastics/glass	√	√	√
Membranes/filters	√	√	√
Tubing	√	√	
IV bags/catheters			√
Syringes			√
Stoppers			√

sure that placebo of acceptable quality and adequate blinding properties is available.

III. MANUFACTURING OF DRUG PRODUCT: ASEPTIC PROCESS CONSIDERATIONS

A. Terminal Sterilization

The FDA requires manufacturers of drugs administered parenterally to provide data to support the necessity of aseptic processing. If a product can be auto-claved, it must be. If the manufacturer claims a product cannot be terminally sterilized, there must be data to support this claim. In the case of RMP-7 drug product, several vials of product were subjected to normal autoclave cycles to determine whether aseptic processing could be supported. The results of this study indicated that an unacceptable level of degradation occurred during even reduced autoclave cycle times (1, 3, 10, 15, 20, and 30 min). This study provided enough evidence to support aseptic processing as the only acceptable means of sterilizing RMP-7 final drug product.

B. Process Equipment Leachates

In filling low concentration (2 μg/mL) products, highly sensitive HPLC meth-ods are capable of detecting and quantitating very low level process/formula-tion contaminants. Guidelines for impurities in peptide drug products allow for 1% or less of single impurities (ICH guideline). There is the potential during processing and formulation stages to leach into final drug product for-mulations low level contaminants that can be detected in final filled drug prod-uct as impurities. One should take care in testing all potential product contact equipment/containers for potential leachates. It is not always obvious which individual pieces of process/formulation equipment come in contact with drug product during filling operations. Caution should be exercised in determining which pieces of process equipment will contact drug formulations. One should thoroughly investigate this potential problem area before embarking on the formulation of low concentration drug products. All process and formulation equipment, should be thoroughly cleaned and rinsed, and the rinsate solutions analyzed for leachates, before active drug product is processed. This precau-tion could save a great deal of time, money, and lost product, in addition to avoiding a lengthy investigation.

C. Incompatibilities Between Process and Product

In addition to monitoring the product during the autoclaving process, it is important to bear in mind that the container closure system can be adversely affected by the terminal sterilization process. Several publications have addressed the issue of extractables from container closure systems and the associated problems of these leaching into a product. For example, leaching of zinc salts from various rubber closures has been reported to result in precipitation, discoloration, and contamination of products (22). Such leaching processes can be accelerated at elevated temperature, as would be encountered during the sterilization process.

An example of this was encountered during the aseptic filling and terminal sterilization of a phosphate buffered saline solution that was being manufactured for use as a placebo. During visual inspection of the autoclaved product vials, insoluble particulates were observed in the solution and on the gray butyl rubber vial closure. Upon isolation and examination of the particulates by X-ray microanalysis, the insoluble particulates were identified as zinc phosphate. A series of experimental compatibility studies was performed and the source of the particulates was identified as the gray butyl rubber stoppers encountered during autoclaving. These stoppers contained zinc in the form of zinc oxide, and the hypothesis was that zinc was leaching out of the stopper and in the presence of the phosphate buffer yielded water-insoluble zinc phosphate crystals. When Teflon-coated gray butyl rubber stoppers were used in the manufacture and autoclaving of the phosphate buffer, no particulates were generated.

IV. LINK OF FORMULATION BACK TO MANUFACTURING

Well-designed formulation studies yield much knowledge of the inherent degradative pathways to which a particular protein (or peptide) molecule is susceptible. They also reveal under which conditions these degradative processes proceed most rapidly. This in-depth understanding of the molecule's physicochemical stability can aid in explaining problems that might have occurred during purification or how subtle changes in pH or storage and handling conditions might effect the final recovery and integrity of the protein. For example, knowledge of how freeze–thaw cycling affects the protein might aid in making decisions on storage of bulk drug substance at various in-process stages. A protein molecule known (from formulation cavitation studies) to be susceptible to air–water interface denaturation may need to be treated more gently, or a protective excipient may need to be added to the formulation during filtration. A change in the process that reduces air–water interface exposure might be another option in handling the sensitive molecule.

In our experience, changes in the large-scale cell culture process have resulted in changes in the IEF pattern of the purified and formulated protein. Knowledge of deamidation rates under cell culture conditions and exposure times to the media components in the cell culture process allowed prediction of the IEF banding pattern when the process was changed. Such projections are analogous to shelf life projections performed for final products. Modeling of this deamidation process based on solution formulation studies allowed extension of this model to the actual cell culture process conditions. Ultimately the model was used to explain that the changes seen in the purified molecule were only minor charge changes resulting from the extended exposure time of the protein in the modified cell culture process.

V. CONCLUSION

The following step-by-step, how-to- protocol should enable the formulation scientist to proceed through a protein solution formulation development study. While it is important to address all aspects of protein solution formulation, there is a natural sequence to the order in which individual parameters are investigated. Ideally, the protein concentration and solubility are determined first to establish a working range for the initial pH stability study. It is also advisable to assess freeze–thaw stability early in the plan. Once these parameters have been investigated, the initiation of a simple range-finding stability study, as outlined below, is recommended.

A. Step 1 Protocol

1. Protein concentration: 1 and 10 mg/mL
2. pH range in 1-degree increments from 4.0 to 9.0
3. Temperature: 2–8°C only
4. Additional expicients: NaCl to near isotonic
5. Time frame: maximum 3 months
6. Sampling interval: weekly at pH extremes; monthly at remainder

B. Step 2 Protocol

The results of the initial range-finding studies will enable one to refine and narrow the scope of future experiments. The next phase of experiments should focus on a narrower pH and concentration range and should study higher temperatures, possibly adding stabilizing excipients and extending the time frame of the study to at least 1 year, sampling at monthly intervals. Specific buffer ion effects could be investigated during this phase.

The final phase step 3, would involve a series of individual studies to address the remainder of the solution formulation parameters.

C. Step 3 Protocol

1. Autoclave study
2. Photostability study
3. Preservative study (if multidose product)
4. Cavitation/shaking
5. Materials adsorption
6. More in-depth freeze–thaw studies

ACKNOWLEDGEMENTS

The authors gratefully acknowledge the efforts of Antonio Pinho and Gil Olson at Alkermes Inc. In addition, we thank Gerry Bell, Christine Hall, Alison Mares, and Dr. Eugene McNally at Boehringer-Ingelheim Pharmaceuticals Inc.

REFERENCES

1. A. Sillero and J. M. Ribeiro, Isoelectric points of proteins: Theoretical determination. Anal. Biochem., 179:319–325 (1989).
2. M. E. Brewster, M. S. Hora, J. W. Simpkins, and N. Bodor, Use of 2-hydroxy-propyl-β-cyclodextrin as a solubilizing and stabilizing excipient for protein drugs. Pharm. Res. 8:792 (1991).
3. Y. John Wang and Rodney Pearlman, eds., Stability and Characterization of Protein and Peptide Drugs: Case Histories. New York: Plenum Press, 1993.
4. Inactive Ingredient Guide. Washington, DC. Division of Drug Information Resources, U.S. Food and Drug Administration Center for Drug Evaluation and Research, Office of Management, 1996.
5. T. Jossang, J. Feder, and E. Rosenquist, Heat aggregation kinetics of human IgG. J. Chem. Phys. 82(1):574 (1985).

6. E. Rosenquist, T. Jossang, J. Feder, and Ole Harbitz, Characterization of a heat-stable fraction of human IgG. J. Protein Chem. 5(5):323 (1986).
7. Y. Uemura, Dissociation of aggregated IgG and denaturation of monomeric IgG by acid treatment. Tohoku J. Exp. Med. 141:337 (1983).
8. Cheng-Der Yu, Niek Roosdorp, and Shamim Pushpala, Physical stability of a recombinant α_1-antitrypsin injection. Pharm. Res. 5(12): 800 (1988).
9. E. Rosenquist, T. Jossang, and J. Feder, Thermal properties of human IgG. Mol. Immunol. 24(5):495 (1987).
10. M. S. Hora, R. K. Rana, C. L. Wilcox, N. V. Katre, P. Hirtzer, S. N. Wolfe, and J. W. Thomson, Development of a lyophilized formulation of interleukin-2. Dev. Biol. Standard. 74:295 (1992).
11. ICH Harmonized Tripartite Guideline, Quality of Biotechnological Products: Stability Testing of Biotechnological/Biological Products, recommended for adoption at Step 4 of the ICH Process on 30 November 1995 by the ICH Steering Committee: International Conference Harmonization.
12. R. Pearlman and T. Nguyen, Pharmaceutics of protein drugs. J. Pharm. Pharmacol. 44:178–185 (1992).
13. ICH Harmonized Tripartite Guideline, Stability Testing: Photostability Testing of New Drug Substances and Products, Recommended for adoption at Step 4 of the ICH: International Conference Harmonization. Process on 6 November 1996 by the ICH: Steering Committee.
14. The United States Pharmacopeia 23. ⟨1⟩ Injections. The United States Pharmacopeial Convention, Inc., Rockville, MD, 1995, pp. 1650–1652.
15. X. M. Lam, T. W. Patapoff, and T. H. Nguyen, The effect of benzyl alcohol on recombinant human interferon-gamma. Pharm. Res. 14:725–728, (1997).
16. A. H. Batts, C. Marriott, G. P. Martin, C. F. Wood and S. W. Bond, The effect of some preservatives used in nasal preparations on the mucus and ciliary components of mucociliary clearance. J. Pharm. and Pharmacol. 42(3):145 (1990).
17. C. R. W. Beasley, P. Rafferty, and S. T. Holgate, Bronchoconstrictive properties of preservatives in ipratropium bromide (Atrovent) nebuliser solution. Br. Med. J. 294:1197 (1987).
18. R. Dabbah, W. Chang, and M. S. Cooper, The use of preservatives in compendial articles. Pharm. Forum, 22(4) 2696–2704 (1996).
19. Ainley Wade and Paul J Weller, eds., *Handbook of Pharmaceutical Excipients*. Washington DC: American Pharmaceutical Association, and London: Pharmaceutical Press, Royal Pharmaceutical Society of Great Britain, 1994.
20. K. Kakemi, H. Sezaki, E. Arakawa, K. Kimura, and K. Ideda, Interaction of parabens and other pharmaceutical adjuvants with plastics containers. , J. Chem. Pharm. Bull. 19(12):2523 (1971).
21. M. J. Akers, Considerations in selecting antimicrobial preservative agents for parental product development. Pharm. Technol. 8(5):36–44 (1984).
22. Y. J. Wang, and Y. W. Chien, Sterile pharmaceuticals: Packaging, compatibility and stability. Technical Report No. 5, Parenteral Drug Association, Bethesda, MD, 1984.

6

Freeze-Drying Concepts: The Basics

Michael L. Cappola
Boehringer Ingelheim Pharmaceuticals, Inc.
Ridgefield, Connecticut

I. INTRODUCTION

Although the origins of freeze-drying can be traced to early times, commercial development was spurred by military needs, during World War II with the freeze-drying of blood plasma and then in the 1950s by interest in the preservation of food intended for military purposes (1). Today specialized food preparation is still an important application of freeze-drying. However major applications have been developed in the fields of medicine and pharmaceutics, especially in the area of "bioproducts," where macromolecules such as peptides and proteins need to be stabilized and formulated into products for health care.

There are numerous other applications for freeze-drying in many fields. However this is a costly process and therefore is reserved for products of high value.

Ideally, freeze-drying is a process whereby a product in aqueous solution is frozen, producing discrete ice crystals and solute crystals. The solid ice under controlled conditions is sublimed away. Any of the "more tightly bound" water is desorbed by controlled heating. The final product's solute is relatively undisturbed from that originally in solution and is finely divided, with a large surface area.

Since the process takes place at low temperature and the product undergoes dehydration rapidly, degradative processes are significantly reduced, producing a high quality dried product.

II. GENERAL: FREEZE-DRYING PROCESS

The freeze-drying process may be performed at atmospheric conditions or under vacuum. In either case, the following general steps are followed:

1. *Freezing*: the product (whether solution, dispersion, emulsion, or biological tissue) is solidified.
2. *Primary drying (sublimation)*: the "free water" of the product (~70–90%) is removed by sublimation.
3. *Secondary drying (desorption)*: the more "tightly bound" water (~10–30%: e.g., water of hydration, ion–dipole, hydrogen-bonded; water associated with amorphous glassy masses) is removed by desorption.

Figure 1 describes the general mechanism whereby water is removed from a frozen solution during freeze-drying.

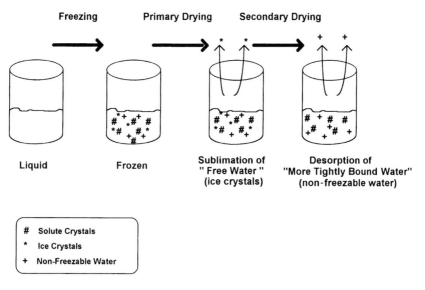

Fig. 1 General freeze-drying process.

A. Freezing

Materials to be freeze-dried can be and usually are complex mixtures of solute in solution that form eutectic and glassy amorphous masses upon freezing. The exact nature of the freezing process and its optimization are product specific. The major issues that need be addressed for specific product freezing are the temperature of complete solidification (including the freezing of any amorphous or glassy materials within the product), along with the optimal rate of cooling, possible thermal treatment, and determination of the presentation temperature for primary drying.

Specific Product Freezing
Major concerns
Temperature of complete solidification Optimal rate of cooling Thermal treatment Presentation temperature for primary drying

Freezing of pure water will start with the formation of ice crystals below 0°C. However the actual formation of solid ice may be delayed by "supercooling" to below −40°C. For ice crystals to form, there must be some seeding or nucleation initiated by solution contents, vibrational energy, or structured water clusters. Lowering temperature contributes to the ordering of water molecules, and therefore low temperatures initiate seeding and final freezing of supercooled pure water (2). Along with the solidification of pure water is an exothermic release of the energy of crystallization, with a resultant rise in temperature. Figure 2 describes these general events in the freezing of a solution containing a eutectic solute.

In the case of a simple eutectic aqueous solution, the solution begins to form ice below its freezing point. As the temperature is lowered (or sustained), more and more water of solution becomes ice, and the solute concentrates; if saturated, it may even precipitate. Significant pH shifts may occur. As the eutectic concentration and temperature are reached, a finely divided crystalline mixture of ice crystals and solute crystals is formed. Figure 3 shows the phase diagram for freezing a simple eutectic aqueous solution. Both the initial ice crystal formation and the eutectic crystallization may be associated with supercooling, and in both instances there is an exothermic heat of crystallization, which raises the overall product temperature before being reduced by further

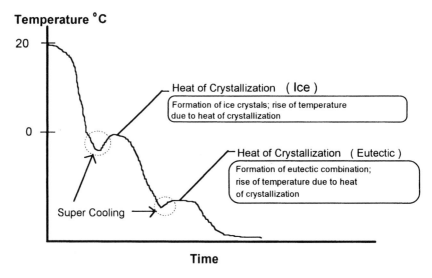

Fig. 2   Freezing of a simple eutectic solution.

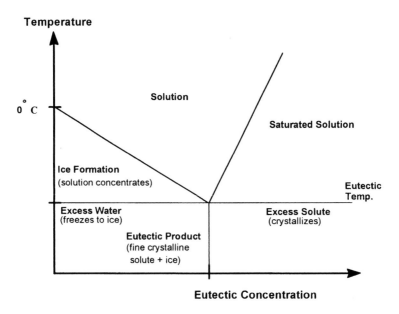

Fig. 3   Phase diagram for a simple eutectic solution.

cooling (Fig. 2). The final temperature of the frozen product in preparation for primary drying will have to be below the eutectic temperature (T_e) to ensure that there is no solubilization of the dried matrix (i.e., "meltback") during sublimation.

With materials such as biological fluids, sugars, polymers, and proteins, which freeze to produce noncrystalline structures, freezing must be aimed at solidifying the product below its glass transition temperature (T_g). The glass transition temperature is characterized by softening of the material upon heating, during which it exhibits fluid characteristics with more molecular disorder and a discontinuous increase in heat capacity. For simple component solutions this is evident via differential scanning calorimetry (DSC) in which heat flow is monitored as a function of temperature (Fig. 4). However, with mixtures of solutes, the T_g becomes more complex and may be difficult to distinguish. Increasing the product temperature above T_g causes softening and viscous flow. At temperatures below T_g, the product is still an amorphous glass, but remains rigid and (for many materials) will be suitable for primary drying. This temperature may also be called the "collapse" temperature (T_c), since above this temperature during primary drying the matrix will be associated with softening, viscous flow, and eventual matrix solubilization or collapse. In an attempt to induce components of a "seemingly" amorphous material to crystallize, the material can be subjected to a thermal treatment during freez-

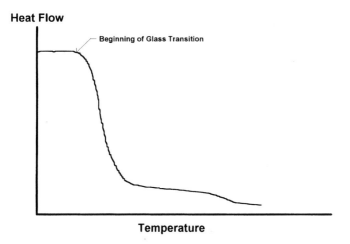

Fig. 4 Glass transition during DSC evaluation.

ing. The product is first frozen and then cycled above and below its T_g in a product-specific way while the formation of solute crystallization is monitored. Figure 5 graphically indicates product temperature cycling during thermal treatment. The advantage that has been noted when this method is used is the decrease in processing time due to an increase in T_c (3). It is of great importance in the development of a product to identify the crystal nature of the product, the desired crystal form to be in the product, and the optimal freezing process.

The freezing process may be evaluated and monitored by a number of techniques, which include differential scanning calorimetry (DSC), differential thermal analysis (DTA), freeze-drying microscopy, and resistivity measurements. DSC and DTA will show the various solution characteristics and crystal transformations with temperature (Fig. 6). The freeze-drying microscope can be used to visually identify and characterize crystal characteristics with cooling rate (Fig. 7 shows the components of the freeze-drying microscope). Resistivity monitoring is useful in the determination of temperature of complete solidification of the product, since resistance usually is at its greatest at this point (Fig. 8).

Freezing of typical materials to be freeze-dried, whether solution, suspension, emulsion, biological specimen, or animal tissue, usually involves complex mixtures of eutectics and substances that will form amorphous

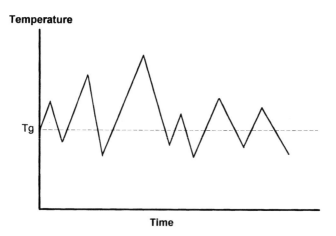

Fig. 5 Thermal treatment of product containing amorphous masses capable of crystallization.

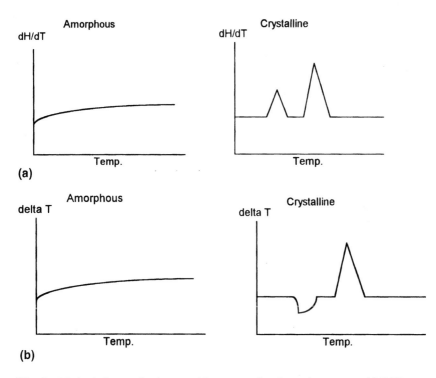

Fig. 6 Methods for monitoring transition states after thermal treatment. (a) Differential scanning calorimetry: Change of heat flow between a sample and an inert reference is plotted against temperature. Sharp changes of heat flow indicate crystal transitions. (b) Differential thermal analysis: Change in temperature between a sample and an inert reference is plotted against temperature. Sharp differences in temperature indicate crystal transitions.

masses. The freezing evaluation will be the combined considerations for that of simple eutectics and that for material producing amorphous structures upon freezing. To prevent matrix meltback/collapse during subsequent drying, the final freezing temperature always must be lower than the lowest temperature for the eutectic (T_e) and amorphous (T_c). Meltback or collapse during sublimation generally compromises product quality and is especially detrimental to biological macromolecules like proteins.

The rate of cooling will affect the degree of supercooling, ice crystal growth, solute concentration effects, pH shifts, and subsequent rate of drying (4). With slow freezing using conventional shelf freezing or prefreezing in a

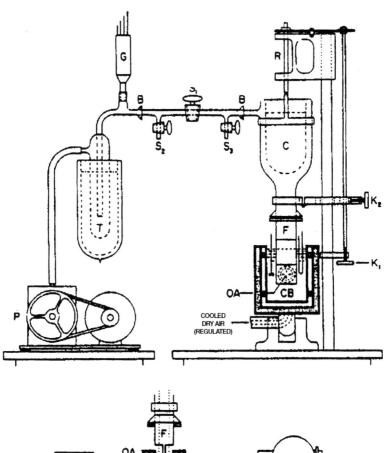

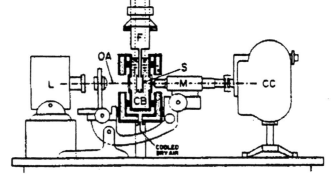

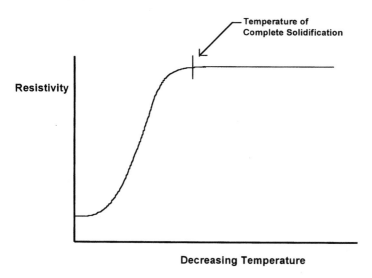

Fig. 8   Resistivity behavior with decreasing temperature for an aqueous solution containing an electrical conducting solute.

freezer ($< 0.2°C/min$), the degree of supercooling is increased relative to a more rapid method. Additionally, the potential for significant shifts in solute concentration and pH is increased. However, since the time for ice crystal formation is longer than a rapid method, ice crystals will grow larger and drying will be faster as a result of the larger pore size. Figure 9 indicates the relative amounts of supercooling for different rates of freezing.

Fig. 7   Schematic views of the freeze-drying microscope. Upper drawing: Front elevation of the apparatus, showing the cooling bath in cross section: B,B, ball joints; C, condenser; CB, cooling bath; F, freeze-drying chamber; G, Pirani gauge; K_1, K_2, stage control knobs; OA, optical axis of system; P, two-stage mechanical pump; R, movable rod; S_1, S_2, S_3, stopcocks; T, refrigerated trap. Lower drawing: Side elevation of part of the apparatus, showing both the freeze-drying chamber and its cooling bath in cross section: CB, cooling bath; CC, cine camera; F, freeze-drying chamber; L, lamp; M, microscope tube; OA, optical axis of the system; S, Sample. (Reprinted from A. P. MacKenzie, Basic principles of freeze-drying for pharmaceuticals, 1965, with permission of Parenteral Drug Association, Inc., Bethesda, MD.)

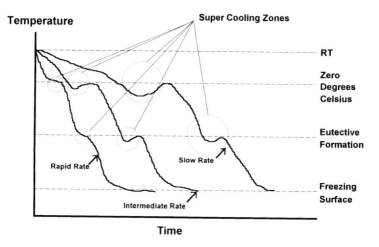

Fig. 9 Relative amount of supercooling based on cooling rates.

Rate of Cooling Affects
_____
1. Degree of supercooling
2. Ice crystal growth
3. Solute concentration effects
4. pH shifts
5. Rate of drying
_____

A rapid rate of cooling (as much as two orders of magnitude greater than the slower method) can be achieved by the immersion method, in which the material to be frozen is quickly immersed in a cold liquid (e.g., liquid N_2, O_2, etc.). Under this method, the material is quickly frozen with a minimum ice crystal size, little supercooling, and a tendency to form amorphous glasses. Solution salt concentration increases and pH shifts are minimized. Although volatile substances may be trapped and therefore better preserved with subsequent drying, ice crystal size is minimum, and drying will be slower because a smaller pore size is generated.

Biological specimens are generally more appropriately treated with the rapid method of freezing, which avoids much of the osmotic and pH damage due to the freeze concentration of solutes during slow freezing. In addition, the use of cryoprotectants [sugars (e.g., sucrose, trehalose); cyclodextrins;

Table 1   Effects of Fast Versus Slow Cooling Rate on the
Freeze-Drying Process

Parameter	Fast cooling	Slow cooling
Degree of supercooling	Small	Great
Solute concentration	Minimal	Maximum
pH shifts	Minimal	Maximum
Ice crystal size	Small	Large
Rate of drying	Slow	Fast

polymers (povidone, polyethylene glycol, dextran); or simple polyalcohols (glycerol)], will stabilize bioproducts against harmful freezing effects by reducing undesirable ice crystal formation and solute concentration effects and increasing amorphous mass formation (5). Table 1 compares the effects of fast and slow cooling.

Generally liquids to be freeze-dried will be filled into containers to a thickness no greater than 10–20 mm. Thickness in excess of this will significantly limit subsequent drying because there will be increased resistance to the removal of frozen water. Similar considerations are made for biological plant or animal tissue unless exceedingly long cycle times are permissible. Conversely frozen product layers of thinner dimensions will accelerate the freeze-drying rate and may be used to offset other formulation and/or process factors that tend to slow the rate. Products may be prepared by a simple fill into a suitable container and subsequent freezing, or they may be spin-frozen on the walls of a vial. Figures 10 and 11 describe several methods of freezing material for drying.

B.   Drying (Primary)

1.   General

After the product has been frozen and any necessary thermal treatments performed, the next consideration is to remove the solidified "free water" without destroying product quality. This is accomplished by primary drying, a sublimation process, which can be performed under either atmospheric or vacuum conditions at low temperature.

Greater than 90% of the product's water may be removed by primary drying, leaving less than 10% to be removed by the next phase of freeze-drying, secondary drying. In either vacuum or atmospheric freeze-drying, heat

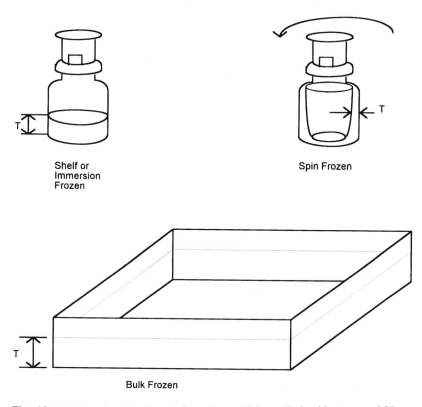

Shelf or
Immersion
Frozen

Spin Frozen

Bulk Frozen

Fig. 10   Solution freezing in containers: layer thickness T should not exceed 20 mm.

energy supplied to the product, while maintaining the product below its lowest eutectic temperature (T_e) or collapse temperature (T_c) for an amorphous material (which we will now call collectively the "collapse temperature" T_c) causes water to sublime from the product. To ensure drying, this water vapor must be removed continuously from the product surface. A safe product temperature setting is generally considered to be 2–5°C below the collapse temperature, to compensate for variations due to equipment. When the product is under vacuum (which is the common method of freeze-drying with pharmaceuticals), a chilled condenser is used to create a driving force for moving water vapor away from the product surface, thereby reducing the partial pressure of water P_w immediately above the product surface (Fig. 12). When atmospheric conditions are used, the water vapor must be mechanically removed

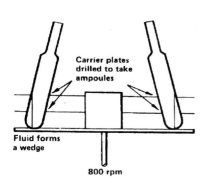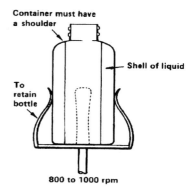

Fig. 11 Formation of wedges and shells of fluid by centrifugation and spinning. (Diagrams courtesy of Edwards High Vacuum, Inc., Tonowanda, NY.)

by increased gas flow above the product surface and/or the use of desiccants, air exchanges, and so on. (Fig. 13).

According to MacKenzie, sublimation may take place by means of four basic mechanisms: (a) direct sublimation drying, (b) direct sublimation via cracks induced in the solute matrix, (c) molecular diffusion through concentrated solute where the matrix structure is preserved, and (d) molecular diffusion through concentrated solute where the matrix structure is not preserved (Fig. 14) (6).

> *Direct sublimation drying* occurs when the frozen material is connected by a continuity of ice crystals such that the sublimation process has a continuous path for the removal of water vapor as the drying process advances through the frozen product. Simple salts (e.g., KCl) that form eutectics are examples of materials that will sublime via this mechanism.
>
> *Direct sublimation via cracks* results with certain proteins and materials that form amorphous masses when frozen. Matrix cracking occurs simultaneously with the freeze-drying process and provides ready channels for the removal of subliming water.
>
> *Molecular diffusion through concentrated solute where the matrix structure is preserved* can occur in concentrated solutions like povidone, gelatin, or sucrose. In these solidified materials the ice crystals are physically separated by solute. The escape of subliming water must include molecular diffusion, possible recondensation, and final sublimation from the product surface.

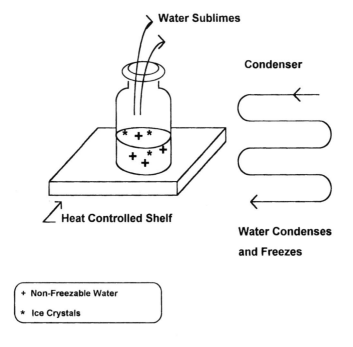

Water Sublimes

Condenser

Heat Controlled Shelf

Water Condenses and Freezes

+ Non-Freezable Water

* Ice Crystals

1. Frozen product is in an evacuated chamber.

2. Heat energy is applied through shelf while maintaining product temperature below "collapse" temperature.

3. Water leaves product and migrates to condenser where it freezes.

Fig. 12 Vacuum freeze-drying.

> *Molecular diffusion through concentrated solute where the matrix is not preserved* occurs when the product temperature exceeds the collapse temperature T_c and the matrix softens, becomes viscous, and appears to be frothing.

2. Heat Input

Heat energy may be applied by conductive, convective, and/or radiant methods (7). The majority of freeze-drying processes in the pharmaceutical industry are performed under vacuum using mainly conductive heating. However, to a certain degree, convective and radiant energy input occurs simultaneously

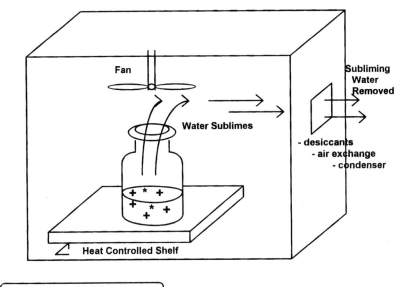

+ **Non-Freezable Water**
* **Ice Crystals**

1. Frozen product is at atmospheric conditions.

2. Partial pressure of water is reduced above the product surface by
    mechanically "sweeping" the area.

3. Heat energy is applied through the shelf while maintaining product
    temperature below "collapse" temperature.

4. Water leaves product and is removed by desiccants, air exchanges,
    condensation, etc.

Fig. 13 Atmospheric freeze-drying.

without being the major contributors to energy flow (Fig. 15). For conductive
heating, heat energy is applied to the bottom of a container, after freezing,
through a heat-controlled shelf.

Convective heating (e.g., via fluidized-bed techniques) of foods at atmospheric conditions has been actively investigated with varying degrees of success (8–11). Product particle size, air velocity, and temperature are the controlling variables. The frozen product is subjected to a controlled air velocity and
temperature at atmospheric conditions, with the sublimed water removed by

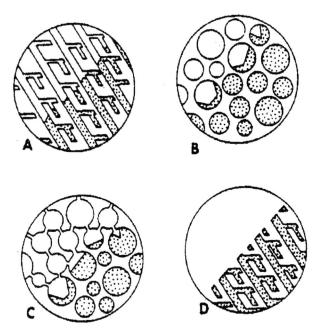

Fig. 14 Mechanisms by which freeze-drying has been observed to proceed in numerous aqueous solutions. (A) Direct sublimation of ice. (B) Water molecules diffuse through freeze-dried matrix solids. (C) Freeze-drying matrix cracks; water vapor passes through fissures. (D) Freeze-drying proceeds with ''collapse'' of the solute matrix. (Reprinted from Ref. 6 by permission of the publisher W. B. Saunders Company, Orlando, FL.)

desiccants, condensation, or air volume replacement. Radiant heating, which has the capability of heating the top and sides of a product during freeze-drying, as of now, is predominantly used in the food industry. Microwave heating techniques have been investigated but have no commercial importance at present (12).

In the pharmaceutical industry containers are placed on a thermal conducting shelf, on which they are first frozen (may be prefrozen by another method, then placed onto the shelf). Then heat is applied under vacuum, such that the product temperature is kept below the collapse temperature as the free water is sublimed to a chilled condenser. Freeze-drying initially occurs at the top surface of the vial and then progresses deeper and deeper into the vial.

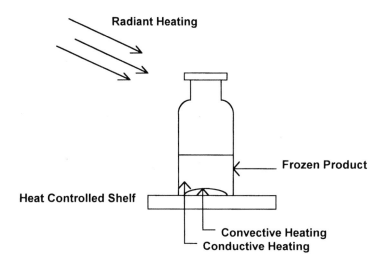

1. Radiant heating from light energy (photon sources).

2. Conductive heating through a heat controlled shelf.

3. Convective heating from a heated gas in contact with the heat controlled shelf.

Fig. 15 Sources of heat input.

With the resultant increased depth of the dried material, there is an increased resistance to further sublimation of the frozen material. Additionally, the product temperature is reduced by surface cooling as water is sublimed from the product surface. Therefore, several activities occur during progression of the primary drying process:

> resistance to heat transfer to the product by physical barriers (shelf material, vial bottom, thickness of frozen layer)
> surface product temperature decrease due to energy loss to sublimed water
> initial rapid rate of sublimation succeeded by a progressively slower rate due to an increase in the dry product thickness

When heat is initially applied to the frozen product under vacuum, the overall rate of sublimation is limited by heat transfer through the shelf, vial, and frozen

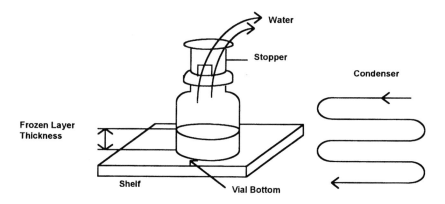

Fig. 16 Barriers to heat transfer.

product (Fig. 16). The sublimed water, under the appropriate vacuum, leaves the product surface and migrates to the condenser surface. The condenser is sized to handle an ice quantity for the maximum freeze-dryer load and still provide the efficiency necessary for rapid production. Heat energy can be applied to offset the surface cooling effects and raise the product temperature close to the collapse temperature for the most efficient process. As the process progresses and there is a consequent increase in dry product thickness, the overall resistance to the escape of the subliming water is increased, and no longer is the process controlled by heat transfer through the different resistances; rather, it is now limited by the mass transfer of ice away from the frozen layer (13,14). However, the major resistance is generally due to the thickness of the dry product (Fig. 17) (15,16). There are other resistances to the flow of water to the condenser, such as those due to vial geometry, stopper design, condenser design, and placement. To prevent "collapse" of the product, the heat input may have to be reduced as the process progresses.

The heat input to the product must be adjusted, taking into account the surface cooling effects of sublimation and the barriers to heat and mass transfer. To achieve economical operation, the process should be adjusted to proceed as close to the collapse temperature as possible without exceeding it. One way to accomplish this is to monitor product temperature (via product temperature probes) while visually observing the product undergoing freeze-

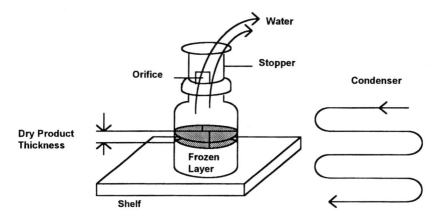

Fig. 17 Barriers to mass transfer of water away from product.

drying. In this way product temperature can be correlated with visual collapse. During developmental experiments, shelf heating can be varied at different time intervals during the cycle to prevent collapse and maximize process efficiency. These changes can then be programmed into a ''proposed'' cycle and the process confirmed by additional test cycles and by product stability testing. Table 2 lists heat input considerations related to barriers to heat and mass transfer.

3. Condenser

A typical process involves loading vials of drug solution into a commercial shelf freeze-dryer, freezing the product (and thermally treating it if necessary),

Table 2 Heat Input Considerations

Barriers to heat transfer	Barriers to mass transfer
Product frozen layer thickness	Dry product thickness
Product vial type and uniformity of contact surfaces	Container size, orifice opening, stopper design
Efficiency of heat transfer of shelf (construction quality and material)	Condenser placement

cooling the condenser to its lowest temperature, and applying the appropriate vacuum, followed by heating the product for water removal. The condenser is set to attain the lowest temperature, such that the vapor pressure of ice in the product at the process operating temperature is significantly greater (by several orders of magnitude) than that of ice on the condenser surface. This vapor pressure difference provides a driving force for the migration of water vapor away from the product surface and onto the condenser, where it is removed as ice after the process. As ice accumulates on the condenser, the surface warms, thereby reducing condensation rate. Therefore, modern freeze-dryer design includes condenser sizing, which minimizes condensation rate differences due to ice loading. The condensers associated with commercial freeze-dryers have the capability of attaining temperatures of approximately -40 to $-95°C$.

Sequence of Freeze-Drying Events

---
1. Product frozen (optional thermal treatment)
2. Condenser cooled
3. Vacuum applied
4. Heat treatment program

---

4. Vacuum

Vacuum is applied during freeze-drying of pharmaceutical products to increase the vapor pressure of water in the frozen product. For example at 100 μmHg, 1 g of water will have a 10 million-fold increase in volume over that at atmospheric conditions (17). Vacuum level may be classified as follows (18):

> Low: $(7.5 \times 10^4)–(1.0 \times 10^3)$ μm
> Medium: $1.0 \times 10^3–1$ μm
> High: $1–1.0 \times 10^{-5}$ μm[2]

Vacuum levels in a freeze-dryer are measured mainly by three types of means:

1. *Hydrostatic gauges*: These gauges measure directly the actual pressure exerted by the gas in the system and include Bourdon, capsule, and McLeod gauges, and U-tube manometers.
2. *Thermal energy gauges*: Pressure measurements are made based on the thermal conductivity of a gas; thermocouples and Pirani gauges are in this group (19).

3. *Pressure transducers*: Pressure measuremnets are made by measuring the deflection of a diaphragm with an electric transducer (e.g., capacitance manometer) (20).

Using a high level of vacuum during primary drying in an attempt to provide the maximum vapor pressure for sublimation is not the best procedure, since the thermal conductance of heat energy into the product at high vacuum levels is decreased and can be a limiting factor. Additionally, an extremely high vacuum level may lead to product contamination due to extraction of volatiles from rubber components inside the chamber or "back-streaming" of oil from the vacuum pump (21,22). However, a "safe" vacuum level can be optimized to provide enough vacuum for good thermal conductance and yet still provide a level high enough to ensure suitable conditions for a high vapor pressure of water in the frozen product. Usually this involves setting the vacuum level at 10–30% of the vapor pressure of ice in the frozen product (23). Additionally, dry nitrogen or other suitable gases can be bled into the chamber in a cyclic fashion, which allows the chamber pressure to vary between a relatively high level and a low level, resulting in a more complete heat transfer at the high pressure and a better sublimation rate at the lower pressure. This process has been used successfully in the food and pharmaceutical industries and takes advantage of both competing processes, providing a faster process than a noncycled process (24).

The end of primary drying is signaled when the product units (vials, containers, etc.) in the chamber all attain the same temperature that will be approximated by the shelf temperature. Additionally, since the water vapor pressure in the chamber is essentially the total pressure during primary drying, a steady decrease in partial pressure of water (detected by a moisture analyzer or gas analyzer) is another indicator of the end of primary drying (25). Because of the number of units in the chamber and the inherent variability between them, usually a period of several hours delay is factored into the process to bring all the units to the same point prior to starting secondary drying.

C. Drying (Secondary)

As indicated at the outset, secondary drying is a desorption process (featuring the removal of unfrozen water that is more "tightly bound" (e.g., water of hydration, ion–dipole, hydrogen bonded; water associated with amorphous glassy masses, etc.) than the free water removed during primary drying. Usually the shelf temperature is raised to the highest temperature possible consistent with the product stability, and the vacuum level increased (pressure de-

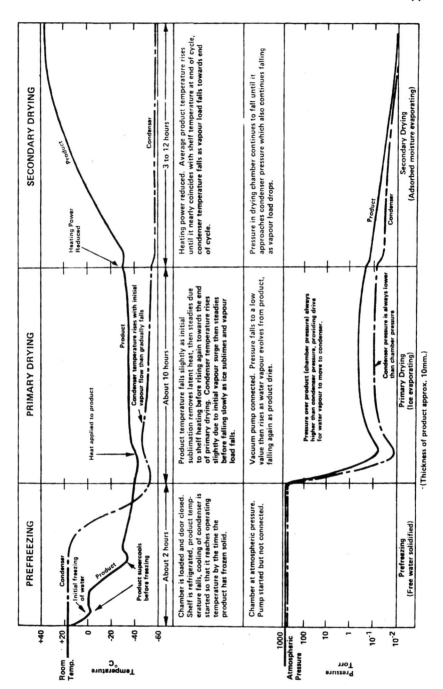

Fig. 18 Chart of complete freeze-drying cycle. (Courtesy of Edwards High Vacuum, Inc., Tonawanda, NY.)

Table 3 Vacuum Considerations

Monitoring
  Hydrostatic gauges
  Thermal energy gauges
  Pressure transducers
During primary drying
  Set vacuum level to 10–30% vapor pressure of water in the frozen material at
    the primary drying temperature or
  Cycle vacuum between "high" and "low" levels consistent with physical
    stability
During secondary drying
  Product temperature is increased consistent with product stability, vacuum level
    is increased to ~200 μm

creased) to "boil off" the remaining water. Chamber pressure need not be reduced too low, however, because of the potential problems mentioned earlier, and it has been noted that levels below 200 μm do not increase significantly the rate of drying (26). Figure 18 shows the product conditions during a typical freeze-drying run, and Table 3 lists vacuum considerations for both primary and secondary drying.

The end point of secondary drying should be determined during early product development by in-process sampling with gravimetric (gravimetric Karl Fisher, thermogravimetric) and/or mass spectrometric analysis of the samples. Desorption isotherms should be determined for the secondary drying time (27). The product stability is then evaluated at different residual moisture levels and optimized for the best performance. Residual moisture levels below 1% w/w are attainable, and the general product range is 0.5–4% w/w.

III. FORMULATION DEVELOPMENT

A. Excipients

The most common functions of excipients in the formulation of freeze-dried products are as follows:

  as bulking agents
  for pH and osmotic adjustment
  for cryoprotection

for product stabilization
for adjustment of "collapse" temperature

Bulking agents or matrix builders provide the structure of the freeze-dried product and are of particular importance when the active ingredient is in low concentration ($< 1\%$). At these low concentrations the active ingredient may, to a certain degree, move out of the product container and migrate with the subliming water to the condenser (28). The matrix in these instances provides the mass to hold the product, thereby avoiding potency loss due to the processing conditions and providing a pharmaceutically elegant product in which the product has little or no volume shrinkage. Usually solution solids are adjusted between 2 and 30%. Above 30% solids, the overall rate of water removal is slowed down significantly because of the additional resistance from the high solids load. Common excipients used for bulking agents in the pharmaceutical industry are mannitol, glycine, lactose, and sucrose. Dextran, although more hydroscopic and not widely used, is a pharmaceutically acceptable bulking agent. These agents combine a relatively high collapse temperature while producing a pharmaceutically elegant matrix or cake. In addition to their matrix abilities, bulking agents may impart useful qualities in regard to increasing the rate of solution, modifying collapse temperature, providing freeze/thaw protection, and/or increasing shelf stability.

Common Bulking Agents
Mannitol
Lactose
Glycine
Sucrose
Dextran

Various buffers, which include the phosphates, acetates, and carbonates, lactates, ascorbates, and citrates, are used to control pH. As mentioned earlier, since there will be concentration, solidification, and crystallization of buffer components during freezing, there will be potential shifts of pH at high buffer concentrations. Therefore, buffers should be kept at low concentration if pH shifts and solute concentration during freezing are damaging to the product.

Salts (e.g., NaCl, KCl, etc.) are used as tonicity modifiers to control osmotic pressure. Osmotic adjustment for parenteral solutions ideally should be to ~ 280–295 milliosmol/liter (physiologic tonicity). However, deviations from isotonicity may be acceptable when injection volumes are small and the infusion rate is slow (29). These salts should also be viewed as potential dehydrating agents, since they concentrate during freezing, creating an unfavorable environment for certain products such as proteins.

Cyroprotectants are excipients that stabilize the product during the freeze-drying process. This may involve protection offered during:

the initial freeze/thermal treatment
and/or primary drying
and/or secondary drying

Cryoprotectants can offer stabilization against harmful salt concentration and pH shifts during freezing by closely associating with the active moiety. During drying, these materials can increase the product glass transition temperature (T_g) and thereby increase process stability. In addition cryoprotectants may stabilize the product by minimizing overdrying during secondary drying (30). Substances that may be important for formulations as cryoprotectants include polymers such as povidone, polyethylene glycol, and dextran; sugars such as sucrose, glucose, and lactose; amino acids such as glycine and arginine; and human serum albumin (31–33).

Product stabilizers (unlike cryoprotectants, which ensure process stability) provide long-term product shelf stability. In reality, cryoprotectants and stabilizers may share the same functions; that is, cryoprotectants may offer long-term stability benefits and stabilizers may be good cryoprotectants.

Excipients are used as ''collapse'' temperature modifiers when, for example, a drug substance with a very low collapse temperature is combined with a bulking agent (or combination of excipients) having a relatively high collapse temperature. The formulation will then have some intermediate collapse temperature depending on the nature of the product, the bulking agent, and the relative amount of each. Figure 19 shows how collapse temperature can be modified by changing the weight ratio of excipients.

The majority of solutions for freeze-drying are strictly aqueous. However solvents, especially ethanol, have been used in mixed aqueous solutions for solubility or extraction purposes. In these cases a trap (liquid nitrogen) is placed before the vacuum pump to remove the noncondensable vapor to ensure that it does not contaminate the vacuum pump oil. Figure 20 shows the design of several cold traps.

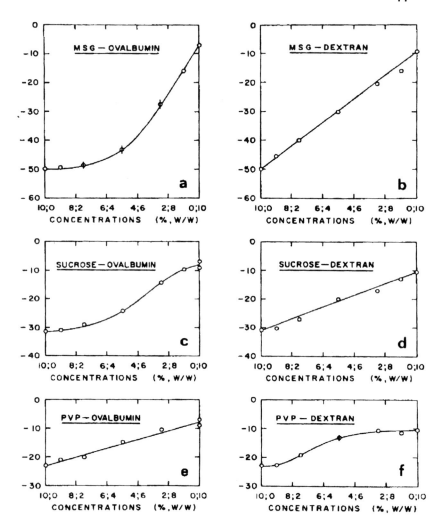

Fig. 19 Dependence of collapse temperature on the composition in six typical three component (two-solute) systems. (Reprinted from A. P. MacKenzie, Collapse during freeze-drying: qualitative and quantitative aspects, In: Freeze-Drying and Advanced Food Technology, Academic Press, London, 1975, p. 288, with permission of the publisher.)

Solid and liquid gas refrigerants reduce the investment in a mechanical refrigerator but for daily use the latter saves money after about 6 months' operation

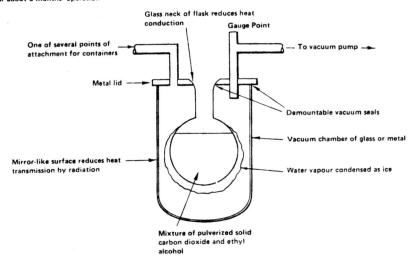

PREFERRED FORM OF COLD TRAP using solid refrigerant. May be modified to take liquid nitrogen. A 5 litre round flask containing maximum quantity of CO_2 lasts 36 to 40 hours under vacuum without condensing duty. Demountable arrangement facilitates defrosting of condenser and removal of water

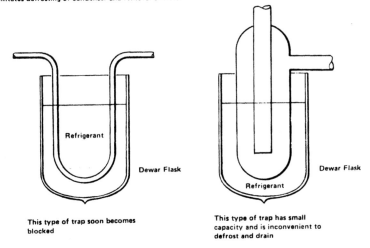

Fig. 20   Cold traps: recommended and nonrecommended forms. (Courtesy of Edwards High Vacuum, Inc., Tonawanda, NY.)

B.   Containers

Vials and ampules comprise the major containers for small-volume freeze drying. Bulk containers of glass and stainless steel are also commonly used (Fig. 21). The U.S. Pharmacopeia standard for glass intended to contain parenteral solutions is USP type I clear or amber borosilicate, however USP types II and III may also be suitable under certain conditions (34).

Vials are available molded or extruded with thin or thick walls and bottoms. All containers intended for freeze-drying should have uniform thickness on the heat transfer contact surfaces (bottom and sides of the containers), and

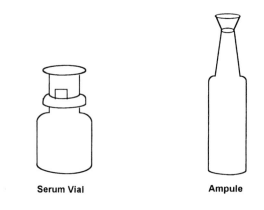

Serum Vial   Ampule

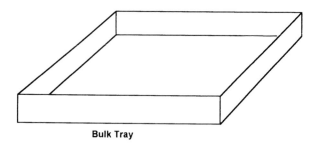

Bulk Tray

Fig. 21 Containers for freeze-drying.

the overall wall thickness should be relatively thin, with enough strength to provide unit integrity. In general, molded vials do not have these characteristics and should not be used. Figure 22 describes some common problems associated with molded vials. Extruded glass vials and ampules are available with uniform thin walls.

Stoppers of different configurations are used to seal vials and even some designs of ampules under vacuum (Figs. 23 and 24). Butyl rubber is the elastomeric material that is commonly used in stoppers because of its low moisture permeability. However other materials that may be suitable are also available (35,36). These stoppers can be coated with a silicone oil or other FDA-approved coatings to facilitate stoppering and to ensure a good seal (37). Crimped aluminum seals are used to securely fix stoppers in place.

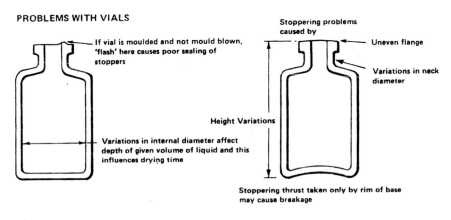

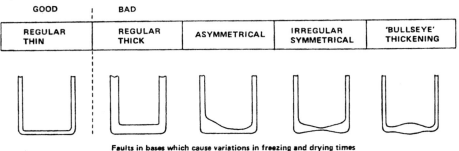

Fig. 22 Problems with vials. (Courtesy of Edwards High Vacuum, Inc., Tonawanda, NY.)

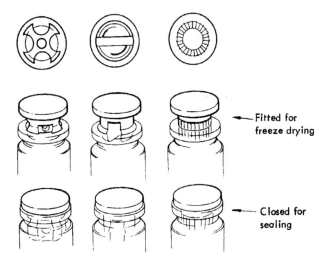

Fig. 23 Stoppers of various types. (Courtesy of Edwards High Vacuum, Inc., Tono-wanda, NY.)

C. Formula Considerations

1. General

The initial formula development must determine the following (38):

> selection of formula excipients and concentration based on preformula-tion studies
>
> solution pH
>
> selection of dose/unit and solution volume
>
> selection of product container/closure and whether the product will be stoppered under vacuum
>
> determination of formula collapse temperature and secondary drying temperature
>
> determination of primary and secondary drying times
>
> determination of residual moisture or loss on drying (LOD) consistent with product stability

The formula ingredients will include the drug, usually a bulking agent, and, if necessary, buffers, salts, cryoprotectants, and stabilizers. The total solids content is typically 2–30% and may be adjusted with a bulking agent to form an elegant cake. The desired solution concentration and dose/unit will deter-

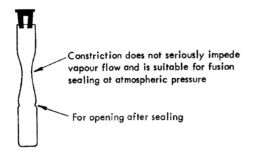

Fig. 24   Preconstricted vampule. (Courtesy of Edwards High Vacuum, Inc., Tono-wanda, NY.)

mine solution volume, which will then provide the information for the selection of the appropriate container size. Containers should be sized such that the frozen product thickness is no greater than 2.0 cm. This product thickness provides a conservative and effective freeze-drying time that will not be greatly affected by resistance to vapor flow.

To establish preliminary process parameters, a determination of the "collapse" temperature of the product formulation must be established. This can readily be accomplished with the use of a freeze-drying microscope (Fig. 7) in which a sample of the product is placed and viewed. The sample is first frozen at low temperature ($< -50°C$ or as low as possible), the condenser is then cooled, and vacuum is applied. When the vacuum level is established, the product temperature is raised slowly ($0.5-1°C/min$ or slower) until there is visual collapse. The product temperature is then lowered until there is retention of the matrix again. Any crystallization changes are noted. The "collapse" temperature is verified by viewing several samples and observing repeatable collapse.

Alternately, a trial-and-error approach in a small freeze-dryer may be used in which a small number of samples are frozen and then, under vacuum/condenser, the product temperature is raised until the product is visually collapsing (frothing). Again the product temperature is lowered such that no frothing is visible. The ramping up and down should be done a number of times with several samples to bracket the "collapse" temperature. Finally a DSC method relating product glass transition temperature(s) T_g to "collapse" temperature may be developed. Once the "collapse" temperature has been determined, the actual temperature used in a commercial freeze-dryer should be set $2-5°C$ lower (to compensate for variability due to differences relating to the shelves, vials, sensing devices, etc). Table 4 gives sample values for the formation and freeze-drying of a typical bioproduct.

Preliminary estimation of primary drying time can be made using the "rule of thumb" that 1 mm of frozen material will take approximately 1 hour to sublimate its free water. Therefore, 15 mm of frozen product will take approximately 15 hours of primary drying time. Quantification of the end of primary drying can be determined as mentioned on page 181. For secondary drying, a good understanding of the product's thermal stability is imperative, since during secondary drying the product temperature should be raised as high as possible consistent with product stability. Product residual moisture or loss on drying (LOD) must be determined during development and final product quality confirmed by appropriate stability studies. The processing time

Table 4   Example of Formula and Process for a Monoclonal Antibody

A. Selection of formula
   1. Concentration of active ingredient
      Product concentration of 2.7 mg/mL
   2. Solution pH
      pH 6 determined to have best product stability: pre-freeze-drying solution,
        dry product, reconstituted solution
   3. Formula excipients
      Bulking agent: 3% w/v mannitol
      pH adjustment: 1% w/v NaH_2PO_4
      Cryoprotectant and stabilizer: 0.75% w/v glycine
B. Selection of dose/container
      To deliver 5 mg/vial (2 mL solution containing 5.4 mg drug)
C. Product container
      5 mL extruded clear type I borosilicate glass serum vial with siliconized gray
        butyl rubber stopper, stoppered under vacuum/nitrogen atmosphere
D. Temperature and time
   1. Collapse: approximately $-7°C$
   2. Product primary drying
      $(-10°C)$ for 10 hours
      $(-10°C$ ramp to 25°C); 2.5 hours
      Total 12.5 hours
   3. Secondary drying
      (25°C) for 4.5 hours
F. Residual moisture
      2–4% w/w

to establish this residual level can be determined by thiefing time samples during experimental process work and determining their moisture content or LOD.

2. Proteins

Proteins offer additional challenges during freeze-drying owing to the sensitivity of the protein structure. In addition to the freezing precautions already mentioned in regard to pH shifts and freeze concentration of solute salts, the freeze-drying process may result in protein degradation. Among the many degradative processes that may occur are dimerization, aggregation, denaturation, deamidation, hydrolysis, oxidation, disulfide exchange, isomerization, racemization, and imide formation (39). Obviously these processes must be avoided or minimized to produce an acceptable product. Protein destabilization due to freeze/thaw (freeze/thermal treatment) can usually be controlled by optimization of formula excipients, especially buffer salts, along with selection of a suitable cryoprotectant. Care must be taken to avoid overheating the product during primary drying, for this would cause "collapse" and possible thermal degradation. Proper cycle heating can prevent such overheating. The elevation of product temperature for secondary drying and its duration are sources of potential danger. Secondary drying of protein products should be aimed at the minimum temperature and time needed to efficiently produce a stable product. Residual moisture levels must be determined consistent with the best product shelf life. The use of cryoprotectants and/or stabilizers should be used to provide process and/or long-term stability. Assessment of final product quality and stability evaluation of protein products may include the following:

1. Physical appearance
   a. Appearance, including color, uniformity, volume (dry product)
   b. Rate of solution (reconstitution time)
   c. Clarity of solution, size, and amount of particulates (reconstituted solution) (as determined by laser light scattering techniques)
2. Preservation of the native protein structure
   a. High pressure liquid chromatography techniques of size exclusion (SEC), hydrophobic interactions (HIC), and ion exchange separations
   b. Gel electrophoresis techniques of native-PAGE, SDS-PAGE, isoelectric focusing, and gel blotting

3. Activity assays
    a. Enzyme-linked immunosorbent assay (ELISA)
    b. Cell-based assay
    c. Enzymatic activity assay

IV. EQUIPMENT

A. Freeze-Dryers

A number of freeze dryer types have been used:

manifold
chamber
fluidized-bed
chamber spray atomization
continuous process

Of the freeze-dryers above, the ones of major importance in the pharmaceutical industry are the manifold type (for research purposes) and the chamber type (for product development and production). The other types are used in the food industry or have been researched for potential application in specialized areas.

The manifold systems are usually reserved for small-scale experimental work, where temperature control of the product during drying may not be of critical importance. The physical appearance of this system is shown in Fig. 25. A container is first frozen, and then the container top is attached to an evacuated manifold. Within the evacuated area is a chilled condenser to collect the subliming water. The container temperature; may not be controlled during drying except by ambient temperature; however, there are manifold systems that offer cycle temperature control. Stoppering under vacuum is generally not possible.

The chamber freeze-dryer is the pharmaceutical industry standard piece of equipment for freeze-drying. It consists of heat-controlled shelves for freezing and heating, a chamber that can be evacuated, and a condenser coil that can be located inside the chamber or externally. With freeze-dryers having remotely located condensers, a valve can be installed between the chamber and condenser for isolating the chamber from the condenser for pressure analysis. The shelves of the freeze-dryer are capable of coming together, such that stoppers on top of the vials in the dryer are forced into the vials, thereby stoppering the vials under vacuum. Different designs for moving the shelves together for stoppering have been devised, including hydraulic and screw-

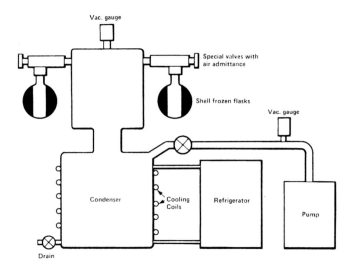

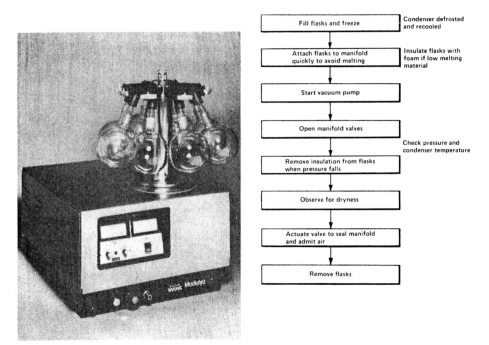

Fig. 25   Sequence for manifold freeze-drying. (Courtesy of Edwards High Vacuum Inc., Tonawanda, NY.)

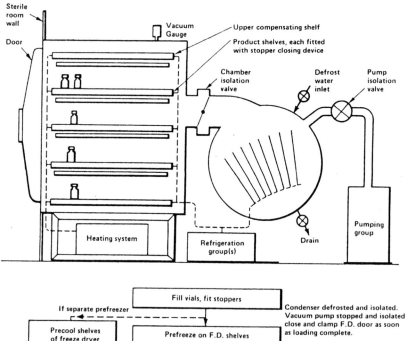

Fig. 26 Sequence for shelf freeze-drying. (Courtesy of Edwards High Vacuum, Inc., Tonawanda, NY.)

actuated. The freeze-dryer for processing pharmaceuticals must be compliant with good manufacturing practices (GMP) in that it must be constructed of 316 stainless steel (or equivalent) in product contact areas and, if intended for processing sterile products, must be sterilizable by steam or chemical means. Also, moving parts must not shed particles. Modern freeze-dryers with cascade refrigeration are capable of shelf temperatures of −70°C and condenser temperatures of −95°C (Fig. 26).

Product temperature may be monitored by probes, which include thermocouples, resistance temperature devices (RTDs), and thermistors (40). Resistivity probes are also available for process evaluation (Fig. 27). Modern systems are computer-interfaced for complete cycle automation and data collection.

B. Vacuum Pumps

Mechanical rotary vacuum pumps comprise the most common means of generating vacuum for freeze-drying (Fig. 28). These pumps are available in single

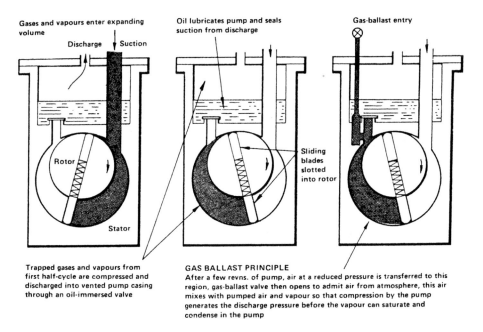

Fig. 28 Mode of operation of simple oil-sealed rotary pump. (Courtesy of Edwards High Vacuum, Inc., Tonawanda, NY.)

A. Product Temperature Probes

Thermocouple Wire - A temperature measuring sensor that consists of two dissimilar
                     metals joined together at one end (junction) that produces a
                     small thermoelectric voltage when the junction's temperature is changed.

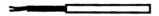

Resistance Temperature Device (RTD) - A probe containing an element that operates
                                       on the principle that electrical wire resistance
                                       is a function of temperature.

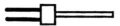

Thermistor -   A probe that has a linear output voltage in response to changes
                in temperature.

B. Resistivity Probes

Resistivity Probe - Conductivity of the solution is determined by measuring the
                     resistance to current flow in the solution.

Fig. 27   Probes used for freeze-drying.

stage (containing one pumping device) and double stage (containing two pumping devices) and are capable of an ultimate vacuum of 10^{-2} torr (single stage) or 10^{-4} torr (double stage) (41). A "gas ballasting" feature provides a means whereby limited amounts of moisture entering the pump can be emitted into the exhaust without condensing in the oil. However systems using organic solvents should be fitted with a cold trap (e.g., liquid nitrogen) before the vacuum pump to contain "noncondensable" vapors. Other pumps such as mechanical booster and entrainment vacuum pumps are also available for freeze-drying; however these pumps are usually reserved for special applications (42).

ACKNOWLEDGMENT

The author gratefully acknowledges George Gereg, AS, of Boehringer Ingelheim Corporation for his preparation of the majority of figures in this chapter.

REFERENCES

1. C. J. King, Freeze-Drying of Foods. Cleveland, CRC Press, 1971, p. 10.
2. H. Rupprecht, Basic physio-chemical principles of freeze-drying—Lyophilization. Farm. Vestn. 44:196 (1993).
3. C. J. King, Freeze-Drying of Foods. Cleveland, CRC Press, 1971, pp. 35–46.
4. H. Rupprecht, Basic physio-chemical principles of freeze-drying—Lyophilization. Farm. Vestn. 44:197–206 (1993).
5. H. Rupprecht, Basic physio-chemical principles of freeze-drying—Lyophilization. Farm. Vestn. 44:201 (1993).
6. A. P. MacKenzie, Principles of freeze-drying. Transplant. Proc. VIII(2), suppl. 1:183–186 (June 1976).
7. J. D. Mellor, Fundamentals of Freeze-Drying. New York, Academic Press, 1978, p. 71.
8. O. Boeh-Ocansey, Effects of vacuum and atmospheric freeze-drying on the quality of shrimp, turkey flesh and carrot samples. J. Food Sci. 49:1457–1461 (1984).
9. O. Boeh-Ocansey, A study of the freeze-drying of some liquid foods in vacuo and at atmospheric pressure. Drying Technol. 2(3):389–405 (1983–84).
10. H. Gilbert, Method of and apparatus for freeze-drying previously frozen products. U.S. Patent Appl. 866,012 (December 1977).
11. G. J. Malecki, P. Shinde, A. I. Morgan, Jr. and D. F. Farhas, Atmospheric fluidized bed freeze-drying. Food Technol. 24:93–95 (May 1970).
12. C. J. King, Freeze-Drying of Foods. Cleveland: CRC Press, 1971, p. 66.

13. J. W. Snowman, Formulation and cycle development for lyophilization: First steps. Pharm. Eng. November/December 1993, p. 27.

14. J. D. Mellor, Fundamentals of Freeze-Drying. New York: Academic Press, 1978, pp. 52–56, 130–160.

15. M. J. Pikal, Freeze-drying of proteins, I. Process design. Biopharm. September 1990, pp. 19–24.

16. A. S. Munns and M. Airah, Freeze-drying with cyclic pressure. Aust. Refrig. Air Condi. Heat, March 1972, pp. 18–19.

17. T. W. G. Rowe and J. W. Snowman, Edwards Freeze-Drying Handbook. Crawley, England: Edwards High Vacuum, 1978, p. 23.

18. N. S. Harris, Vacuum Technology. Crawley, England: Edwards High Vacuum, 1977, p. 7.

19. T. W. G. Rowe and J. W. Snowman, Edwards Freeze-Drying Handbook. Crawley, England: Edwards High Vacuum, 1978, pp. 11–17.

20. J. Armstrong, Use of the capacitance manometer gauge in vacuum freeze-drying. J. Parenter. Drug Assoc. 34, (6):473–483 (November–December 1980).

21. J. Y. Lee, GMP compliance for lyophilization of parenterals. Part I. Pharm. Technol. October 1988, p. 58.

22. N. A. Williams and G. P. Polli, The lyophilization of pharmaceuticals: A literature review. J. Parenter. Sci. Technol. 1984, pp. 54, 57.

23. M. J. Pikal, Freeze-drying of proteins. I. Process design. Biopharm. September 1990, p. 24.

24. A. S. Munns, and M. Airah, Freeze-drying with cyclic pressure. Aust. Refrig. Air Condit. Heat. March 1972, pp. 18–22.

25. M. L. Roy and M. Pikal, Process control in freeze drying: Determination of the end point of sublimation drying by an electronic moisture sensor. J. Parenter. Sci. Technol. 43(2):60–66 (March–April 1989).

26. M. J. Pikal, Freeze-drying of proteins. I. Process design. Biopharm. September 1990, p. 24.

27. J. C. May, R. M. Wheeler, N. Etz, and A. Del Grosso, Measurement of final container residual moisture in freeze-dried biological products. International Symposium on Biological Product Freeze-Drying and Formulation. Dev. Biol. Stand. 74:153–164 (1991).

28. M. J. Pikal, Freeze-drying of proteins. II. Formula selection. Biopharm. October 1990, p. 28.

29. I. Reich and R. Schnaare, Tonicity, osmoticity, osmolality, osmolarity. In: Remington, ed. The Science and Practice of Pharmacy, 19th ed., Vol. 1. Easton, PA: Mack Publishing Co., 1995, p. 615.

30. M. J. Pikal, Freeze-drying of proteins. II. Formula selection. Biopharm. October 1990, pp. 28–29.

31. M. J. Pikal, Freeze-drying of proteins. II. Formula selection. Biopharm. October 1990, pp. 28–29.

32. M. E. Ressing, W. Jiskoot, H. Talsma, C. W. van Ingen, E. C. Beuvery and D. J. A. Crommelin, The influence of sucrose, dextran, and hydroxylpropyl-β-

cyclodextrin as lyoprotectants for freeze-dried mouse IgG_{2a} monoclonal antibody (MN12). Pharm. Res. 9(2):262–270 (1992).

33. T. Arakawa, Y. Kita, and J. Carpenter, Protein–solvent interactions in pharmaceutical formulations. Pharm. Res. 8(3):285–291 (1991).

34. Task Force No. 17, Parenteral Drug Association, Glass containers for small volume parenteral products: Factors for selection and test methods for identification. Philadelphia, PA. Technical Methods Bulletin No. 3. Parenteral Drug Association, Inc., (1982).

35. T. W. G. Rowe and J. W. Snowman, Edwards Freeze-Drying Handbook. Crawley, England: Edwards High Vacuum, 1978, p. 63.

36. Task Force No. 14, Parenteral Drug Association, Elastomeric closures: Evaluation of significant performance and identity characteristics. Philadelphia, PA. Technical Methods Bulletin No. 2. Parenteral Drug Association, Inc., 1981.

37. Lubrication of Packaging Components Task Force, Siliconization of parenteral drug packaging. Technical Report No. 12, J. Parenter. Sci. Technol. suppl. 42(48) (1988).

38. J. W. Snowman, Formulation and cycle development for lyophilization: First steps. Pharm. Eng. November–December 1993, pp. 28–34.

39. Y. J. Wang and M. A. Hanson, Parenteral formulations of proteins and peptides: Stability and stabilizers, Technical Report No. 10. J. Parenter. Sci. Technol. suppl. 42(2S):4–8 (1988).

40. Omega Complete Temperature Measurement Handbook and Encyclopedia, Vol. 28. Stamford, CT: Omega Engineering, 1992, pp. A-5, C-4, D-6.

41. N. S. Harris, Vacuum Technology. Crawley England: Edwards High Vacuum, 1977, pp. 23–25.

42. T. W. G. Rowe and J. W. Snowman, Edwards Freeze-Drying Handbook. Crawley, England: Edwards High Vacuum, 1978, p. 29.

7
Formulation of Proteins for Pulmonary Delivery

Andrew R. Clark
Inhale Therapeutic Systems
San Carlos, California

Steven J. Shire
Genentech, Inc.
South San Francisco, California

I. INTRODUCTION

Formulation of proteins and peptides often is more challenging than formulation of small molecules because of the important role of protein conformation as well as the potential for numerous chemical degradation pathways (1,2). This fact coupled with the necessity of using a device to generate aerosols augments the challenge considerably. Although the developed formulation must provide 1–2 years of stability on storage it also must meet additional requirements that are unique to its delivery as an aerosol. First, the formulation must not cause adverse pulmonary reactions such as cough or bronchoconstriction. Secondly, the formulation components should not interfere with the generation of the aerosol. For example, the amount of product that is generated as an aerosol should not be decreased, and the desired size distribution should not be altered by formulation components. In particular some devices such as nebulizers will involve exposure of the protein to large surface areas, and the formulation may have to be designed to minimize interactions with the component materials. Finally, the formulation also must stabilize the protein

sufficiently to ensure that the protein survives the rigors of the aerosol generation process. In addition to these challenges, the development and ultimate
approval of an aerosol formulation often goes hand in hand with development
and/or use of a particular device.

II. EVALUATION OF DEVICES AND FORMULATION COMPATIBILITY

A. Aerosol Parameters and Device Performance That Impact Drug Delivery

The delivery efficiency of a device/formulation system will rely essentially
on two key performance parameters: the efficiency of aerosol production and
the generation of suitable particles or droplets for aerosol deposition into the
airways. The efficiency E of an aerosol generation system is a measure of
how much of the drug product is actually converted into an aerosol, as opposed
to how much is actually retained by the device, that is

$$E = \frac{\text{aerosol mass}}{\text{total loaded mass}} \tag{1}$$

The deposition of aerosol particles or droplets in the airways is a function of
both size and density. Aerosol particles encounter passageways of smaller
diameter as they enter the deep lung periphery, and thus the size of the particles
can determine how far the particles are able to travel. Although the size is
undoubtedly important, the mass of the particles also plays a large role in
airway deposition. In particular, particles with greater mass tend to impact on
upper airway surfaces and are not deposited in the lower regions of the lung.
Thus, the relevant parameter governing aerosol deposition appears to be the
aerodynamic diameter d_{aer} which is related to the physical sphere diameter d
and particle density ρ by (3):

$$d_{aer} = d\rho^{1/2} \tag{2}$$

It has been shown that for polydisperse liquid aerosols, droplets with mass
median aerodynamic diameters larger than 5–6 µm deposit in the oropharyn-
gal region during normal tidal breathing, whereas droplets smaller than 1 µm
tend to be exhaled without significant deposition (4,5). Although somewhat

arbitrary, the portion of the size spectrum between 1 and 6 μm is generally considered to contain the bulk of the particles with the potential to penetrate and deposit in the lungs, and is referred to as the fine particle fraction (FPF)*. This is an obvious oversimplification, since the size distribution desired will be different depending on which region of the lung is being targeted for delivery. In particular, if the target is the alveolar region, it would be more appropriate to generate a size distribution on the lower end of the 1–6 μm region. Since the amount of aerosol deposited in the lung will depend on both the efficiency of aerosol generation and the fine particle fraction, an overall delivery efficiency, DE can be defined as the product of these parameters:

$$DE = E \times \text{FPR} \tag{3}$$

Clearly the actual deposition of aerosol will depend on additional factors in individual patients (breathing pattern, lung anatomy, pulmonary obstruction, etc.) However, in vitro comparisons of devices/formulations in terms of DE can guide development of a formulation and choice of device.

B. Characterization of Aerosols

There are essentially four key questions that need to be answered in regard to delivery performance of a device:

1. How much of the drug is converted to an aerosol that ultimately exits from the mouthpiece?
2. What is the size distribution of particles or droplets in the aerosol?
3. What is the reproducibility of the aerosol generation process?
4. What effect does the device/or aerosolization process have on the protein drug quality?

---

* The fraction of the aerosol with this size range was referred to in the past as the "respirable fraction," but recently it has been agreed to refer to it as the "fine particle fraction" (International Society for Aerosols in Medicine Focused Symposium: Towards Meaningful Laboratory Tests for Evaluation of Pharmaceutical Aerosols, Puerto Rico, Ja. 29–31, 1997; Proceedings in the Journal of Aerosol Medicine, Volume 11, 1998). Often the range of sized chosen for the FPF is operational and is dependent on the chosen device for sizing the aerosol particles that is, the cutoff values for the stages of an impactor or liquid impinger will define the exact range used by the investigator.

The first three questions require methods to analyze the size distribution in the aerosol and drug deposition within the device and any connecting conduits (if applicable) to the size measuring apparatus. Although a variety of methods have been developed to characterize aerosol size distributions, including laser diffraction (6), holography (7), static (8) and dynamic light scattering (9), and time-of-flight aerosol beam spectrometry (10), with the exception of the use of laser diffraction for the assessment of the size distribution of nebulizer clouds (11), cascade impaction (12) and impinger methods (13) are considered to be the most reliable method for the assessment of aerosol particle size distribution, and device performance (14). The advantages of the alternate methods include their speed and ease of analysis in comparison to impactor and impinger technologies.

A great advantage of the impinger and impactor technology, however, is that it readily allows for a determination of aerosol mass balance. This is particularly critical when one is addressing device efficiency and size distribution. Clearly if a majority of the aerosol is made up of large particles that are not collected because of impaction onto surfaces prior to entry into the measuring device, then the distribution will be skewed to lower sizes. Similarly small particles that are not collected or recorded by the measuring device will lead to a distribution skewed to larger sizes. Since the target is the human airways, it is reasonable to use an artificial throat that will collect the larger size particles that normally are deposited in the oropharyngeal region. Similarly, a filter can be placed after the size detector to trap the smallest particles/droplets that escape detection. Determination of the reproducibility of the dose delivered can be a challenging exercise, especially in the case of devices that deliver small amounts of drug. The limitations of such an analysis will depend on the sensitivity of the assays used to detect the protein, as well as the ability to recover the aerosol reproducibly.

The characterization of protein that has been aerosolized also requires the collection of most of the protein exiting the aeorsol generation device. Often this is done by impaction, but small particles/droplets are difficult to collect by impaction. A successful characterization of the protein drug will, of course, require a good set of stability-indicating assays, but the full recovery of protein for analysis is critical to ensure that protein contained in the small particles/droplest (<1 μm diameter) has not undergone any degradation. A device and technique for increasing the size of nebulized droplets and the surface area for impaction has been developed and was successful in collecting ~96% of the aerosol (15).

III. PROTEIN FORMULATIONS FOR AEROSOL DELIVERY

Formulations of protein for aerosol delivery can be developed as either liquid or solid dosage forms, as in the case of formulations for parenteral administration. The development of solid dosage forms for aerosol adminisration has its own unique set of challenges and is discussed later in this chapter. The following section discusses the development of liquid formulations for aerosol delivery.

IV. PROTEIN LIQUID FORMULATIONS FOR AEROSOL DELIVERY

A. Choice of Device

As already mentioned, the approval of a formulation often is linked to the device used for generation of an aerosol. The method used to generate the aerosol will dictate what components are required to ensure protein stability. A common device for generation of aerosols is the so-called jet nebulizer (Fig. 1a). A portable air compressor generates a high velocity airstream through a jet nozzle, and liquid is drawn up from the reservoir as a result of the partial pressure drop at the orifice. A droplet spray is generated upon contact with the airstream, and the larger droplets impact on an appropriately placed baffle, ultimately returning to the liquid reservoir. Droplets of sufficiently small size remain in the airstream and exit from the inhalation port of the nebulizer. The result is that during the course of nebulization more than 99% of the solution is essentially refluxed and undergoes repeated stress and exposure to air–water interfaces, an experience that may promote protein denaturation (16,17).

Another device for nebulization is the ultrasonic generator (Fig. 1b) which transmits high frequency sound waves through the reservoir solution to generate aerosol droplets. The transfer of high energy as well as the potential build up of heat in the solution may lead to degradation of a protein drug via thermal rather than surface exposure (18). However, the direct generation of small droplets by ultrasonic nebulization has led to the development of devices, such as the Respimat®,* which will generate respirable aerosols with

---

* The original Respimat® was discontinued, and a mechanical device formerly known as the BINEB®, discussed later in the chapter, is now referred to as Respimat.

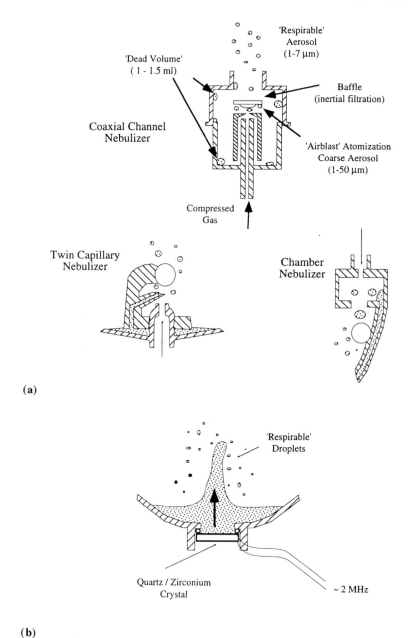

Fig. 1 Nebulizer designs: (a) conventional pneumatic (jet) and (b) ultrasonic.

a small volume of solution, typically 25–50 μL (19). This circumvents the potential degradation that may occur upon recirculation of the solution within the nebulizer, but it necessitates the development of formulations that support a higher concentration of protein. As an example, 2.5 mL of Pulmozyme® at 1 mg/mL is delivered by jet nebulizers with an estimated delivery efficiency of 10% (20). {Pulozyme is a recombinant human deoxyribonuclease (rhDNase) used in the treatment of cystic fibrosis.] To deliver a similar amount of drug to the lung by means of a small-volume ultrasonic nebulizer with equivalent efficiency would require a Pulmozyme concentration of 50 mg/mL. The efficiency of these devices is related to aerodynamic diameter of the droplets generated as well as to the fraction of the nominal loading dose that is converted into an aerosol (20). Generation of an aerosol with a high fine particle fraction leads to greater overall delivery efficiency, which may allow for a lower dosing concentration.

One alternate strategy for greater delivery efficiency while avoiding the rigors of a continuous nebulization or potential degradation due to high energy input by ultrasonics is to generate the aerosol mecanically. Such a strategy is used with the AER$_x$ (Fig. 2) and BINEB® (Fig. 3) aerosol generators (21,22). In the AER$_x$ device, a blister pack containing 25–100 μL of drug is rapidly compressed with a mechanical piston, forcing the solution through a series of laser drilled holes (1–3 μm in diameter). An advantage of such a system is the unit dose packaging. Other systems that use a reservoir to supply small volumes to an aerosol generator require development of a multidose formula-

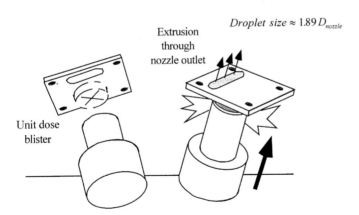

Fig. 2   The AER$_x$™ concept, a multinozzle unit dose liquid atomizer. (From Ref. 21, with permission.)

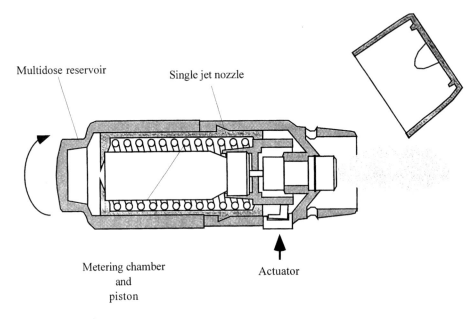

Fig. 3   The BINEB®, a novel multidose liquid atomizer. (From Ref. 22, with permission.)

tion. Multidose formulations of protein, in turn, call for the use of a preservative which may interact with the protein, either reducing the effectiveness of the preservative or affecting protein activity. The BINEB also uses a mechanical piston to generate the aerosols but is actually designed as a metered dose inhaler (23). The drug is stored in a collapsible plastic bag, and metered quantities are delivered to the atomization chamber. Nonreturn valves prevent backflow and control the flow of liquid through the device. However, as with most devices of the multidose liquid reservoir type, it is probably essential that the formulations be preserved.

A further development in liquid atomizers is the ultrasonic mesh. In such devices solution is atomized as it is forced to flow through tiny holes in a vibrating plate. Recent developments have shown that it is possible to produce droplets that are approximately half the diameter of the mesh hole size (24), and this method appears to have much potential. However, the issues with regard to unit dose packaging or preserved solutions still apply, and to date this technology has been applied only to handheld nebulizers.

B. Choice of Excipients

The osmolality and the pH of solutions are critical variables that may affect bronchoconstriction and contribute to adverse reactions during pulmonary delivery of drugs (25–30). It has also been recommended that, whenever possible, solutions for pulmonary delivery, especially those delivered as large volumes by nebulizers, be formulated as isotonic solutions at pH values exceeding 5 (27). Recent studies have shown that if the formulation is not isotonic then the droplet size distribution of an aerosol may be altered during delivery as the result of a loss or uptake of water vapor from the airways (31,32). It is also not uncommon to find that buffer components can cause adverse reactions such as cough (33,34) and, therefore, many inhalation products have been formulated without buffer components. The control of the pH in an unbuffered formulation is a major concern, especially since many protein degradation pathways are highly pH dependent (2). However, at high enough protein concentration the titratable amino acid residues of the protein may provide sufficient buffering capacity to stabilize the pH of the formulation (35).

One of the biggest challenges in developing a liquid formulation for aerosol delivery is the exposure of the protein to an air–water interface. As described earlier, the generation of respirable droplets greatly increases the protein exposure to this interface. Solvent–protein interactions are critical for maintaining the native conformation of a protein, and removal of the aqueous phase can have profound effects on protein structure. In particular, the unique properties of water, such as its ability to form an extensive hydrogen bond network, are believed to be essential for the entropically driven hydrophobic forces that play a major role in folding of proteins (36). Hydrophobic amino acid residues tend to organize water structure by forming cavities in the bulk solvent, leading to a large decrease in the entropy of the system. Removal of the hydrophobic residues from the solvent phase and coalescence into an interior hydrophobic phase results in a decrease of the protein surface area, as well as an increase in the entropy of the solvent phase large enough to exceed the configurational entropic decrease due to folding of the protein into a more compact form.

A protein exposed to an air–water interface during nebulization may become denatured, forming both soluble and insoluble aggreagate (37). Often inclusion of an acceptable surfactant such as polysorbate 20 can minimize this degradation (Fig. 4). The grade of polysorbate may be critical in this application. In particular, polysorbates have been shown to contain trace quantities of peroxide (38). The protein is being stressed by the aerosol generation process, especially in the case of jet and large-volume ultrasonic nebulization, and

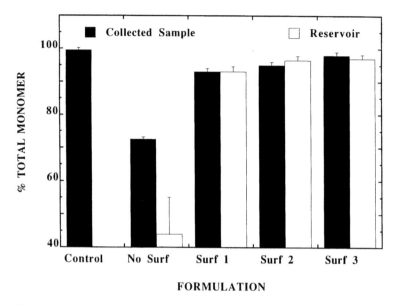

Fig. 4 Effect of surfactant on human growth hormone (hGH) aggregation following 18 minute nebulization of 25 mg/mL hGH. Unaerosolized control (left), collected sample by impaction into test tube (■), sample remaining in reservoir (❑). (From Ref. 37, with permission.)

exposure to large quantities of oxygen may lead to oxidation of susceptible amino acid residuces such as cysteine, methionine, tryptophan, and histidine (2,17). Contamination with peroxides may increase the rate of oxidation, which may destabilize the protein and lead to an increase in aggreagation during aerosolization. The increased potential for oxidation is not necessarily a problem. Whereas many proteins have altered conformation or altered activity, or undergo aggregation, there are also many examples of oxidation having no perceptible effect on the protein (2).

If oxidation does pose a problem, excipients such as antioxidants or amino acids such as methionine can be added as oxygen scavengers. However, the addition of these excipients may cause adverse events in clinical situations, depending on the sensitivity of the patient population. This, in fact, is generally the major limitation on formulation development of proteins for pulmonary delivery. The problem is particularly difficult if the targeted population involves pulmonary diseases such as cystic fibrosis or asthma. The potential sensitivity of asthmatics to excipients makes the development of a multiuse formulation even more problematic than in the development of parenteral formulations. Many aerosol generators that have been developed use an open system, with a reservoir that will require the addition of a preservative to the formulation.

Interaction of the preservatives with the protein is a major concern. If the interactions are substantial, protein denaturation and possibly soluble aggregate formation of precipitation may result (39). The interactions of protein with preservatives may also result in a decrease in the effectiveness of the preservative. This presents significant challenges, especially if the protein drug is being developed for world markets, since the preservative challenge requirements in Europe are significantly more stringent than in the United States. A variety of preservatives delivered as aerosols have been tested in people. Among these preservatives are benzalkonium chloride, chlorbutol, benzyl alcohol, sodium metabisulphite, chlorocresol, and phenol, and some of them have been implicated in adverse effects such as bronchoconstriction (40–44).

C. Manufacturing Issues

As already discussed, the development of a formulation for aerosol administration of a protein drug depends highly on the type of device that will be used. The concern regarding adverse events from excipients or preservatives in patients with compromised lung function may be minimized by the use of devices that administer small volumes. The trade-off, of course, is that the protein may need to be formulated at higher concentrations. The ease of accomplishing this will be related to the solubility and tendency of the protein to aggregate. In addition, although the target concentrations, perhaps as high as 50 mg/mL, may be attainable, the long-term stability of the product may be compromised at the higher concentrations. In particular, bulk product will need to be stored prior to loading (into cartridges, blister packs etc). for use with the device. If long term storage is required, it may be necessary to freeze formulated bulk. If, however, the developed liquid formulation lacks the necessary stability to undergo repeated freeze/thaws cycles, the formulation may have to be stored at controlled temperatures of 2–8°C. Generally, bulk formulations in large quantities are stored in stainless steel tanks, and this practice poses several additional challenges. In particular, a useful pharmaceutical tonicifier such as sodium chloride may cause problems in long term storage because the well-known interactions of stainless steel with halides can result in metal catalyzed oxidation of proteins (45). In such a case, alternate tonicifiers may have to be explored. Sugars such as mannitol are pharmaceutically acceptable, but there may be a risk of adverse events such as bronchoconstriction associated with exposure to large amounts of the sugar (46). Despite these reports, it should be noted that there are no published toxicology data that suggest there is a long-term safety concern on the use of mannitol or other sugars in respirable products. Indeed, lactose has been used for many years with no reported problems. Moreover, any concerns may be less problematic for formulations for small-volume aerosol delivery devices, since the total amount of excipients delivered is quite low in the small volumes (<50 µL) used.

Additional complications may occur if the formulation is designed for multiple uses and hence requires preservatives. This has been demonstrated in studies of the effect of metals and the preservative phenol on human growth hormone degradation (47). Although a single-use formulation does not require preservatives, compliance with recent FDA proposals nevertheless requires sterility for liquid formulations intended for inhalation (48).

If the pharmaceutical company also possesses the technology to make and fill the devices, the required storage time from formulation to final packaging may be minimized. Unfortunately, the pharmaceutical company developing a product for inhalation therapy often, does not possess the device technology and is thus compelled to form an alliance with a particular company that has developed a device with the required performance for efficient aerosol delivery. Such an arrangement usually necessitates the development of appro-private manufacturing steps for long-term storage and shipment of the product to the device company for loading into drug reservoirs, cartridges, blister packs, or other delivery-ready forms designed for the device. The shipment of formulated bulk places additional stresses on the product and may require the allocation of resources and improvements to allow for the additional manipulations of the protein drug. Any alterations in formulation will still need to be compatible with the airways and should not interfere with the aerosol generation process.

V. PROTEIN POWDER FORMULATIONS FOR AEROSOL DELIVERY

A. Choice of Device

As described above for liquid systems, the registration and approval of a protein aerosol product is intimately linked to the aerosol generation system chosen for its delivery. Hence the choice of a delivery system is crucial to the success of the product. For the purposes of protein applications, dry powder inhalers may be placed into two major categories: multidose devices, where drug powder is stored in bulk and metered inside the device before inhalation, and unit dose devices, where the drug powder is stored as a premetered dose in an individual storage unit. These two categories can be further subdivided into patient-driven devices, where a patient's inspiratory effort provides the power to disperse the powder, and powered systems, where an external source provides the energy.

The first category, multidose devices, present some major difficulties with protein powders. As described below, one of the general techniques for stabilizing proteins in the solid state is the use of amorphous glasses. The Achilles' heal of amorphous glasses, however, is moisture. The excipients used as solid state stabilizers, when stored in the amorphous state, have the potential to crystallize; and as the moisture content of the powder increases and plasticizes the solid, the probability of crystallization, and hence the probability of destabilization of the formulation, increases. Although it is possible to develop formulations that are reasonably stable at ambient humidities, the general physical instability of amorphous solids severely limits the application of multidose-type dry powder inhalers. In general the choice for protein powders is therefore the unit dose approach.

Figure 5 presents a schematic illustration of the various approaches to dry powder inhaler design, indicating some proprietary devices and some manufacturers. The unit dose approach in terms of device design can come in many forms. These range from the original dry powder inhaler designs such as the Spinhaler and Rotahaler, where the drug is stored in hard gelatin capsules, to the more complex foil blister devices, where the drug is stored as individual doses either singly or on multidose disks or tapes. Essentially the latter devices offer unit dose drug storage with multidose convenience for the patient. In essence, because of the humidity instabilities described above, the foil systems become the packaging of choice for protein inhalation powders.

Again it is possible to develop formulations with ambient humidity stability that probably would allow the use of systems based on conventional gelatin capsules or other simple technologies, but the extra security guaranteed by the foil blister technology is desirable. (It should be noted that for satisfactory operation, gelatin capsule technology requires the capsule to be maintained at close to 50% relative humidity: if a capsule becomes too dry, it may shatter instead of opening, and if it becomes too wet it cannot be opened at all).

The other packaging issue, which is not directly related to powder stability, is microbial contamination and growth. However, to date there have been no published reports to suggest that protein powders are any more susceptible

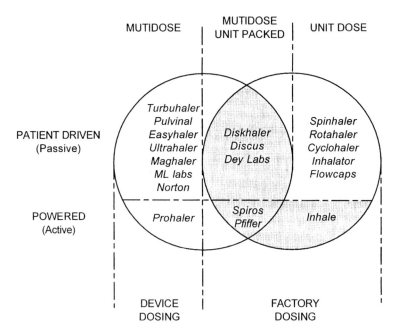

Fig. 5 Current and future dry powder inhalers grouped by dosing method, packaging and dispersion power source. (Shaded area shows area most applicable for protein applications.)

to microbial growth than are the small-molecule powders currently used in commercial inhalation formulations. Both the U.S. and European pharmaco-poeias impose microbial limits for inhalation powders but do not require sterility. Table 1 presents the current proposed microbial limits for powder inhalation products. In general, powders maintained below 50–55% relative humidity will not sustain the growth of organisms; organisms already present may survive in stasis, however.

The remaining choice is then between a patient-driven system (passive) and a powered device (active). Currently only the former type of dry powder system is marketed. However, a number of companies, including Inhale Therapeutic Systems of Palo Alto and Dura Pharmaceuticals of San Diego (49) are in the late stages of development of powered inhaler systems. Figures 6 and 7 show schematics of these devices. The choice between patient-driven and powered systems for protein delivery depends less on the physicochemical characteristics of the molecule, than on the delivery requirements. In general,

Table 1   Assignment of Microbial Limit Tests for Nonsterile Finished Dosage
Forms According to Route of Administration

Route of administration	Total aerobic microbial count (cfu/g or mL)	Combined yeasts and molds count (cfu/g or mL)	Examples of objectionable microorganisms[a]
Inhalations	≤10[b]	≤10	Escherichia coli Pseudomonas aeruginosa Salmonells species
Vaginal	≤100	≤10	Escherichia coli Staphylococcus aureus Pseudomonas aeruginosa Candida albicans
Nasal/Optic/ Rectal/Topical	≤100	≤10	Escherichia coli Staphylococcus aureus Pseudomonas aeruginosa
Oral liquids	≤100	≤10	Escherichia coli Salmonella species
Oral solid	≤1000	≤100	Escherichia coli Salmonella species

[a] It is virtually impossible to list every microorganism that may be objectionable for a specific product class. The microorganisms listed are merely examples of those microorganisms usually found to be objectionable in the respective product class.
[b] Except for nonpressurized powders for oral inhalation for which the total aerobic mircrobial count does not exceed 100 colony-forming units per gram.
Source: Pharmacopoeial Forum, 1996.

patient-driven dry powder inhalers exhibit lung delivery that is both flow rate dependent and less efficient than that obtainable from powered systems. For example, for typical asthma systems, lung delivery may vary by a factor of 3 or more depending on the flow rate achieved by the patient during any particular inhalation and lung deposition is only of the order of 10–20% of the nominal dose. Whereas this set of conditions may not be problematic with some protein therapeutics, where raw material cost and dose consistency are not of concern, it could be disastrous for an expensive protein with a narrow therapeutic window.

Efficiency, E, and fine particle fraction, FPF, are essentially important because of cost. Bioavailability of proteins from the gastrointestinal tract is very poor (50,51). Therefore, while avoidance of oropharyngeal deposition is obviously desirable, it is not as imperative as it is for some small molecules,

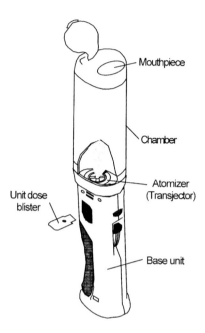

Fig. 6   The Inhale pulmonary delivery system.

where systemic availability from the gut can represent a serious addition to the total unwanted systemic load. In essence, avoiding oropharyngeal deposition by generating a high fine particle fraction results in an overall increase in delivery efficiency, *DE*.

In contrast to the patient-driven passive systems, powered active systems decouple aerosol generation from the patient's inspiratory effort. Hence they reduce the flow rate dependence of the dose delivered to the lung. However, such dependence is not completely eliminated. Since the capture efficiency of the mouth and oropharynx is dependent on the inhaled flow rate, even if aerosol generation is independent of inspiratory flow, the deposition pattern in the body, hence the lung dose, is not. However, this depositional dependency is far less dramatic than the variations that can be brought about by the variations in particle size and flow rate that can be produced by patient-driven dry powder inhalers. A number of authors have used the following function, reported by Rudolph et al. (52), to estimate the oropharyngeal filtering term η_{oral} for monodisperse aerosols:

$$\eta_{oral} = 1 - [1.1 \times 10^{-4}(V_t^{-0.2}\, d_{ae}^2\, Q^{0.6})^{1.4} + 1]^{-1} \qquad (4)$$

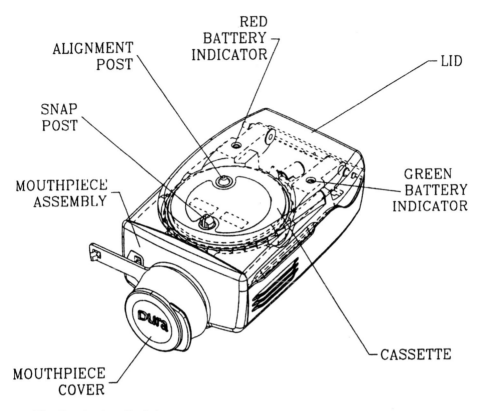

ALIGNMENT
POST

RED
BATTERY
INDICATOR

LID

SNAP
POST

MOUTHPIECE
ASSEMBLY

GREEN
BATTERY
INDICATOR

CASSETTE

MOUTHPIECE
COVER

Fig. 7 The Dura Dryhaler.

where V_t is tidal volume, d_{ae} is the aerodynamic diameter of the inhaled particle, and Q is the inspiratory flow rate. Examination of Eq. 4 shows that as either d_{ae} or Q decreases the probability of penetration beyond the oropharynx and deposition in the lung increases. It should also be noted that at a sufficiently small d_{ae}, Q becomes unimportant and η_{oral} tends to zero.

From a protein delivery perspective then, the most desirable design attributes of stability, delivery efficiency, and reproducibility lead in general to unit dose packaging with a device featuring powered aerosol generation and control of inhalation flow rates. In an imperfect world it may be that all these attributes will not be present in one particular device design, and a number of trade-offs between delivery performance and stability may be necessary. It may also be that for certain molecules, where expense and therapeutic window are not important, simpler, less efficient devices may be acceptable. Indeed, from a patient compliance perspective, they may be even more desirable.

B. Choice of Excipients

As described above, aerosol particles intended to penetrate and deposit in the lungs must be sufficiently fine to pass through the oropharynx and upper airways. In general, this requirement leads to the need for a high fine particle fraction in the delivered aerosol. This in turn leads to the requirement for fine powders of the active drug moiety. In general, fine powders are cohesive and do not flow or disperse easily. The major objective and challenge of powder inhalation formulation is therefore to manufacture powders that both flow, for ease of filling and dispensing, and also disperse, to form aerosols fine enough to penetrate and deposit in the lungs. (The term "fine" generally refers to particles having aerodynamic diameters in the size range 1–6 μm.)

Over the past four decades, two basic approaches have been developed to combine these almost mutually exclusive requirements. The first, developed in the late 1950s, by the then Fisons Pharmaceuticals, is to blend the active drug with a carrier consisting of large particles. This approach produces powders that flow and dispense well and, when a suitable shear force is applied, disperse to produce respirable aerosols of reasonable quality. The coarse carrier used in these systems, typically 30–100 μm in diameter, is usually such that a large fraction of it deposits in the mouth and oropharynx and only a small amount reaches the lung.

The second approach, also developed by Fisons, is to pelletize the drug particle to make loose agglomerates, which flow for filling and dispensing, but which break-up when an aerodynamic shear force is applied to produce a respirable aerosol. A further approach that may show promise is to produce particles of large physical size and low absolute density (53). This approach results in particles that behave as if they have large physical diameter as far as powder flow and dispersion are concerned, but because of their low absolute density possess a suitably small aerodynamic diameter when dispersed (Eq. 4).

The ability of a protein powder to produce a respirable aerosol, whether blended with a coarse carrier or in its "raw" state, is controlled by numerous influences. The diameter of the primary powder particles and their absolute density is of obvious importance. However, the cohesive forces that hold the powder particles together and prevent dispersal of the powder, and the numerous environmental influences that affect these cohesive forces, also play a major role. As an example of how formulation and particle size can influence

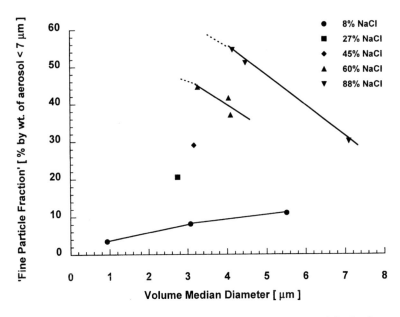

Fig. 8 Summary of the dispersion properties (as FPF) versus particle size for powders containing co-spray-dried NaCl and rhDNase. (From Ref. 54, with permission.)

aerosol performance, Fig. 8 shows data from a study report by Chan et al. (54) detailing the effects of sodium chloride content and powder particle size (expressed as mass median diameter, MMD, of the raw powder) on the aerosol properties of rhDnase powder blends delivered via a Rotahaler. It can be seen that as the sodium chloride content is increased, the FPF increases. That is, the powder disperses more easily and generates a finer aerosol cloud. It can also be seen that the FPF is a function of powder particle size: for the powders that disperse easily, the FPF appears to decrease as MMD increases, whereas for poorly dispersing powders FPF appears to go through a maximum. The effect of particle size is related to cohesion, which is a measure of the force needed to separate a unit mass of particles from their stable agglomerated state. Although the contributions of the various forces to the cohesion of the powder in this study are uncertain, as they are in most studies of this type, if it is generally assumed the van der Waals forces are dominant then the cohesive force would be expected to be proportional to the square of the particle diameter (55). Thus the powders with smaller MMD are more difficult to disperse, and as the MMD increases the powder disperses more easily. How-

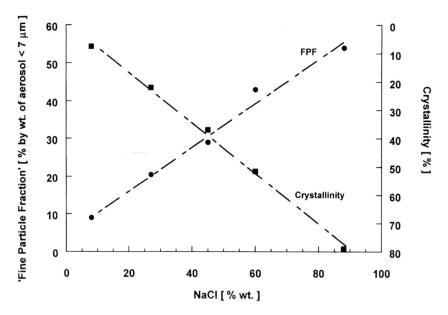

Fig. 9   Relationship between NaCl content, the corresponding crystallinity (pure rhDNase powder is amorphous), and the dispersing properties (as FPF) of rhDNase powders. [All powders have similar primary size distributions before aerosolization, with median diameters of 2.7–3.3 μm (span 1.04–1.63 Mm).] (From Ref. 54, with permission.)

ever, as the primary powder MMD increases, the available amount of fines decreases. Hence, these two competing phenomena result in an optimum powder size for any given formulation, the particular optimum size being dependent on the relative cohesiveness of the powder. Fig. 9 presents data for FPF with a constant powder MMD and varying degrees of crystallinity. For this particular formulation–device combination, an increase in crystallinity results in an increase in the fine particle mass generated from the powder blend. This increase presumably occurs because the interparticulate forces have been modified by the change in physical state of the solid and the external morphology of the powder particles.

In addition to the overall requirements for an inhalation powder formulation, proteins require stabilization in the solid state. The task of manufacturing powders containing proteins that have both the required protein-stabilizing properties and the physical characteristics necessary for generation of fine aerosols can be particularly daunting. Additionally, because in general protein stabilization requires amorphous solids, the pelletization approach, which involves the formation of agglomerates by using small amounts of solvent—which may promote crystallization—can be particularly difficult and is probably unsuited for this application. The formulation approaches thus involve the technical challenges of first manufacturing a fine powder containing the protein and then either working with the fine untreated powder and overcoming the filling and dispensing problems via mechanical means or, as described above, blending the protein-containing powder with a coarse carrier. Most solid state stabilizers used thus far in the development of prototype powder inhalation formulations are those used in stabilization during the lyophilization process. The use of mannitol, lactose, trehalose, and various other excipients has been reported (54,56,57). However, the exact mechanism by which protein stabilization is achieved with these sugars is still in some debate. Theories range from water replacement (58), where the sugar substitutes for the water hydrogen bonds and thus keeps the protein in its native conformation, to mere physical hindrance (59,60), where the protein molecules are simply immobilized in the glassy matrix and are prevented from unfolding and spatially separated, to prevent aggregation. In these solids it is important to achieve both chemical and physical stability.

As described above, amorphous powders, which are desirable from a protein stability point of view, possess the thermodynamic desire to crystallize. Crystallization can lead to the physical degradation of the powder, in terms of flow and dispersibility, as well as to aggregation of the protein (54). Thus, in general, if the physical explanation of stability is to be accepted, the approach is to produce powders with the highest glass transition temperature possible (59). An alternative approach to glass stabilization is to produce powders that contain only small quantities of stabilizing material and are predominantly protein. High protein content appears to prevent crystallization by "poisoning" the crystallization process (61). However, reducing the excipient content of some molecules can increase physical stability, but it may lead to increasing damage to the protein, with aggregation as a result. A delicate balance between physical stability, aerosol performance and protein stability may have to be struck. This is exemplified in the work of Clark et al. (57).

Table 2 presents the effect of lactose on the biochemical stability of spray-dried powders of rhDNase over a 40-week storage period. It can clearly be seen that as lactose content is increased, aggregation decreases. However, Fig. 10 presents FPF as a function of lactose content and humidity over 4 weeks: clearly, as lactose content is increased, the powder performance decreases. Clark and coworkers demonstrated that this decrease in performance is due to crystallization of the lactose. From a stability perspective, then, de-

Table 2 Biochemical Stability Data for Spray-Dried Powders Containing rhDNase and Lactose After Storage for 40 Weeks at 38% Relative Humidity

Formulation (rhDNase:lactose)	Temperature (°C)	Monomer (%)[a]	Δ Deamidation (%)[b]	Relative Activity[c]
100:00	4	97.1	0.3	0.91
	25	93.1	0.6	0.89
84:16	4	99.4	0.5	0.99
	25	97.4	2.0	0.85
66:34	4	99.6	0.1	1.00
	25	98.6	1.5	0.94
50:50	4	99.7	0.1	1.01
	25	99.2	1.6	0.96

[a] Determined by size exclusion chromotography using TSK2000SWXL column. Immediately after spray-drying, samples contained 100% monomer. All values are mean of duplicate determination.
[b] Determined by tentacle cation exchange chromotography using a LiChronsphere 1000 SO_3 column. All values are mean of duplicate determinations.
[c] Determined by rhDNase methyl green assay. Immediately after spray-drying, samples were fully active (i.e., relative activity of 1).
Source: From Ref. 57, used with permission.

signing a protein powder for inhalation may entail the art of choosing a particular compromise between physical and biochemical stability. [In general, the use of lactose as an excipient for protein powders can be contraindicated because of the propensity of this sugar to react with lysine residues, producing lactosylated protein molecules (62).]

Careful assessment of the physical and toxicological acceptability of excipients must also be made. Only a few excipients have Generally Regarded As Safe (GRAS) status from the FDA for use in the lung, and toxicology studies may be required to demonstrate the acceptability of the chosen excipient. Fortunately pulmonary delivery of proteins in general appears to be reasonably safe (63). Also, as with the liquid solution formulations described above, the effects of pH on reconstitution, and induced tonicity changes on dissolution of the powder on the airway walls must all be considered. For example Anderson et al. (46) recently showed that delivering large quantities of sodium chloride, lactose, or mannitol to the airways can induce a broncho-constrictor response in asthmatic patients. Although, even in severe asthmatics lung doses on the order of 5–10 mg had to be delivered before a significant

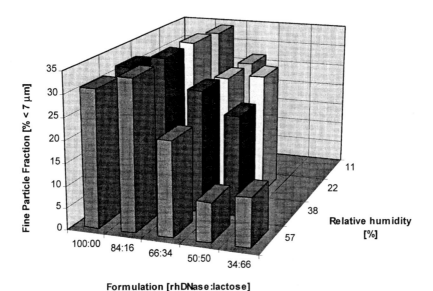

Fig. 10 Fine particle fraction as a function of protein/lactose content and storage relative humidity after 1 month at 25°C. (From Ref. 57, with permission.)

clinical response was seen, the ever wary formulator should keep these issues in mind when developing protein inhalation formulations.

C. Manufacturing Issues

Two basic approaches have been taken in the manufacture of fine protein powders. These involve either lyophilization followed by micronization (64) or spray-drying. Although in general lyophilization of proteins is a well-established process (65), micronization is not, and thus the first approach can be problematic. Conventional jet mills produce local heating due to the particle particle–collisions necessary for comminution and can introduce metallic contamination, which can affect product stability. Although not excluded, this process has had limited success. The second approach, spray-drying aqueous solutions, has been very successful and has been used for many years to prepare dry protein powders. Both these approaches have been reported in the literature. The feasibility of delivering inhalation powders of granulocyte colony stimulating factor produced by spray-drying and combined with coarse carrier lactose has been demonstrated (56). Spray-dried powders containing sodium chloride and rhDNase have been described by Chan et al. (54).

VI. PROTEIN METERED DOSE INHALER FORMULATIONS

In general the metered dose inhaler has not been the delivery system of choice for proteins and peptides. This is presumably because of poor efficiency and reproducibility of delivery, coupled with the relatively harsh propellant environment. However, issues concerning the phaseout of chlorofluorocabons (CFC) propellants and their replacement with hydrofluoroalkanes (HFA) may be involved, as well. The pharmaceutical industry has experienced difficulties in reformulating existing small molecules, and one can only suspect that the formulation of proteins and peptides in HFAs could be even more difficult.

Despite these difficulties, a number of early feasibility studies have been reported. Indeed, a technically successful, although not commercially viable, formulation has been develop for leuprolide, a potent agonist of luteinizing hormone releasing hormone (LH-RH) (leuprolide is discussed as a case study in Sec. VII. B). Brown et al. (66,67) have reported a number of metered dose inhaler preparations involving an antigenic protein, an enzyme, and an antibody. In their first series of experiments they reported two metered dose inhaler formulations based on dimethyl ether (DME). The first contained either alkaline-phosphatase or an IgG_1, κ murine monoclonal antibody. In both cases the proteins were lyophilized with either Laureth-9, Tween 40, or Tween 80 before being added to the DME. In the case of alkaline-phosphatase the lyophilization process alone reduced enzymatic activity fivefold. In case of the antibody, the use of the surfactant appeared to offer protection against denaturation. The metered dose inhalers were shown to produce FPFs of the order of 10–25%. Alkaline-phosphatase appeared to be stable for up to 10 days regardless of the surfactant used, whereas the IgG denatured quite rapidly with Tween-40 or Laureth-9. Their second report (67) describes formulations involving bovine γ-globulin suspended in CFC 11/12 (trichlorofluoromethane/ dichlorodifluoromethane), DME or propane are described. A variety of surfactants were investigated, and FPFs from the MDIs were again around 25%. However, in all cases except Tween 80 and DME, denaturation and a loss of antigenicity occurred rapidly.

These early studies were aimed at utilizing surfactant micelle formation to produce solution aerosols of the protein. To date there have been no reports of attempts to deliver large proteins by means of the powder suspension type of metered dose inhalers. This may be because of the issues of solid state powder stabiltiy described above, or it may be due simply to lack of interest in this dosage form for protein delivery.

VII. PROTEINS FORMULATED FOR PULMONARY DELIVERY AND SPECIFIC CASE HISTORIES

In this section we discuss some case studies in which proteins were formulated for aerosol delivery. The first involves development of a protein for topical delivery to the lung using liquid formulations with nebulizers; the second involves the formulation of a peptide as a metered dose inhaler for systemic delivery. Unfortunately the latter product has had a rather checkered development history and, as a result of the phaseout of CFC propellants, was abandoned as a commercial product. However, because of its unique place in protein/peptide formulation development, it is cited as a case study.

A. Formulation Development and Characterization of Recombinant Human Deoxyribonuclease I (rhDNase) [Pulmozyme®, (Dornase Alpha)]

Cystic fibrosis patients, as a result of persistent bacterial infections, have large concentrations of host-neutrophil-derived DNA in their airway sections. This increased concentration contributes to an increased viscosity of the mucosal secretions in these patients. Recombinant human DNase can alter the viscoelastic properties of human sputum and was, therefore, developed for delivery targeted to the human airways. Pulmozyme was developed as a liquid inhalation product to be delivered by jet nebulizers. Details on the clinical use of DNase (bovine and human), formulation, and characterization of the aerosols are available in a recent review (68). The salient features of that review are presented here.

1. Formulation Development

As discussed earlier, osmolality, pH, and buffering agents can promote adverse reactions in the lungs, especially in patients with compromised pulmonary function. Thus, an isotonic liquid formulation was developed without any buffer components or preservatives. Although buffering agents were not used in Pulmozyme® formulations, it was demonstrated that at 1 mg/mL the protein itself provides sufficient buffering capacity to control the pH of the solution. It was shown that calcium is required for stability of this protein, and thus sufficient $CaCl_2$ was added to maintain activity during storage. Since this formulation did not contain any preservative it was designed for single-use administration in nebulizers.

Many inhalation products are manufactured for single use, using blow-filled seal technology such as that employed by Automatic Liquid Packaging Inc. The plastic vials manufactured by this process are made from low density polyethylene, which is permeable to water vapor and gases. Thus suitable foil packaging also had to be developed for this product. This type of packaging provides great convenience to patients as well as allowing increased throughput for the manufacture of the product. In addition the plastic surface resulted in a lower pH of storage for the product than is obtainable upon storage in glass vials. This actually resulted in longer shelf life for the product, since the major degradation route for Pulmozyme is deamidation of an asparagine residue that results in lower potency of the product.

2. Characterization of rhDNase Aerosols

As already mentioned, the distribution of the size of droplets in an aerosol is critical in determining deposition in the airways. An ideal combination of device and formulation will convert most of the solution into an aerosol that contains a majority of the droplets within the fine particle fraction. Since most cystic fibrosis patients are familiar with the use of jet nebulizers, these devices were used to deliver aerosol, and the in vitro behavior of these systems was characterized by techniques already discussed and reviewed (18,20,68). Initially, a cascade impactor was used to characterize four different nebulizers; the results are summarized in Table 3. The essential findings were that there were no significant differences in performance of the four jet nebulizers (FPF or efficiency, E) and that the overall efficiency of delivery (Eq. 4) was ~25%. This type of performance is similar to that observed for other drugs delivered by jet nebulization. The more rapid analysis via laser light diffraction was later used and gave results comparable to those determined by the more labor-intensive analysis using cascade impactors.

The quality of rhDNase before and after nebulization was determined. It was shown that there was no reduction in activity, no generation of aggregates, and no significant alteration in rhDNase tertiary and secondary structures. Altogether, the four jet nebulizers that were tested were essentially equivalent in their ability to deliver respirable doses of intact, fully active, nonaggregated rhDNase (18,20,35).

Table 3 1 mg/mL rhDNase Delivery by Jet Nebulizers

Nebulizer	n^a	FPF (%)	Nebulizer efficiency (%)	Delivery (%)
Marquest customized Respirigard II	8	51 ± 3	48 ± 6	24 ± 4
Marquest Acorn II	8	50 ± 4	55 ± 6	27 ± 3
Hudson T Up-Draft II	8	51 ± 4	49 ± 5	25 ± 2
BaxterAirlife Misty	7	46 ± 3	44 ± 2	21 ± 1

[a] Table values are a result of n independent measurements using n individual nebulizers, given as the mean value and standard deviation.
Source: Adapted from Ref. 20.

B. Formulation Development and Characterization of LHRH Metered Dose Inhaler

Leuprolide is a 1kDa nanopeptide possessing potent LHRH agonist properties. It is indicated in diseases such as endometriosis and prostate cancer. The molecule is hydrophilic, and it is not absorbed from the gastrointestinal tract. Leuprolide is currently administered by subcutaneous injection. Adjei and Carrigan (69) showed that systemic delivery of leuprolide acetate was possible via the lung, and a series of papers documents various aspects of their metered dose inhaler formulations.

   Proof of concept was demonstrated in beagle dogs using intratracheal administration of 1 mg/mL saline solutions of leuprolide acetate (70). Volumes of solution, based on body weight, were instilled at fixed distances from the epiglottis, and bioavailability was measured as dose-corrected areas under the curve relative either to subcutaneous or intravenous injection. The results of these experiments clearly showed that as the drug was deposited more distally, bioavailability increased: at a distance of 25 cm, corresponding to the bifurcation at the base of the trachea, the drug was essential 100% bioavailable. (*Note*: Leuprolide has both C- and N- terminals blocked, hence is not easily metabolized in the lung.)

   Following proof of concept, human studies were carried out using metered dose inhaler formulations (70). Solution and suspension metered dose aerosols were compared in 23 normal volunteers (69). The solution MDI was formulated using ethanol as the solvent and CFC 12 as the propellant. The suspension aerosol utilized a blend of CFC 11 and CFC 12, the leuprolide acetate being suspended as a micronized powder. All the MDI formulations showed acceptable stability over 3 months. No literature is available to confirm a longer shelf life. The human data confirmed the high bioavailability determined in dogs and demonstrated a correlation between FPF, determined by means of cascade impaction methods (70), and bioavailability. The FPF for the suspension aerosol was four times higher than for the cosolvent, presumably because of the poor evaporation kinetics of the spray (71). Relative bioavailabilities of 6.6% for the solution aerosol and 27.9% for the suspension aerosol were obtained. When corrected for the difference in FPFs, the bioavailabilities were similar: that is, 66% for the solution aerosol and 73% for the suspension.

These data, coupled with the inferred stability, demonstrate the technical feasibility of the MDI approach for leuprolide. However, all the formulations described above used CFC propellants and, as a result of the CFC phaseout, cannot be commercialized. Issues relating to the formulation of leuprolide in HFA propellants have yet to be discussed in the literature.

VIII. CHOOSING BETWEEN DRY POWDER AND LIQUID AEROSOL DELIVERY APPROACHES

This chapter has tried to summarize the merits and problems encountered in developing solid state and liquid formulations of proteins for pulmonary delivery. As already discussed, the difficulties of developing both liquid and powder formulations can be numerous and varied. In the case of powder preparations, formulation development is complicated by the need for biochemical stability of the macromolecule, physical stability of the powder, and a powder that readily lends itself to dispersion and aerosolization. In the case of liquids, the constraints are a little more relaxed, with the major issues being solution stability and survival during a nebulization process. However, the price paid for a solution's early versatility can be a less convenient product, which requires refrigeration during storage and a more cumbersome delivery system.

So where does a formulator begin? As with all development projects, the formulator must begin with the timelines and end product requirements in mind. If the delivery system needs to be readily available for administration when symptoms develop, or if dosing frequency is high, a handheld device coupled with a formulation that is stable at room temperature is highly desirable. Generally liquid formulations of proteins will not be room temperature stable and will not have the stability of a dry powder. Hence, when there is a need for maximum patient convenience, a dry powder product is more desirable. However, powders can sometimes require more ''up front'' formulation development, and there is the major issue of device specificity in formulation development. In fact device availability can be one of the main obstacles to carrying out early clinical work with powder formulations. Currently devices that deliver liquid aerosols (i.e., nebulizers) are readily available, whereas "generic" dry powder devices are not. Also, nebulizer performance is, in general, formulation independent, whereas, powder formulations, again, tend to be device specific. It can therefore be advantageous to begin clinical development and carry out "proof of concept"-type studies with early solution formulations and nebulizer devices, with the intention of utilizing the dry powder approach for later studies and as the commercial product form. With this approach, however, two formulations must be developed, and after "proof of concept," the quickest time to market may sometimes be obtained by continuing development of the solution form. This can lead to an inferior product, but earlier commercial returns.

Although the development of a liquid formulation can often be less challenging, hence quicker, than work involving a dry powder, the requirements for inclusion of a preservative to produce a multiuse formulation may complicate both the development and approval processes. The number of preservatives used in inhalation products is limited and the potential for direct interaction between protein and preservative is always present. It may be possible to limit these problems by selecting appropriate excipients, but this tactic may add considerable time to the formulation development process. The obvious way around these issues is to use a device capable of administering individually packed unit doses. Another approach could be to use multidose systems with nonventing closures. However, even these systems may have to employ preservatives in order to ensure an acceptably low risk of microbial contamination.

It should be clear that there is no single approach to the development of a protein pulmonary delivery product. In general, formulation and device must be coupled early in the development cycle and must be developed as an integrated system. If there is no, or limited, access to a dry powder device, the development of a liquid formulation and the use of nebulizer devices may facilitate early clinical work. However, the ultimate choice between dry powder and liquid must be dictated by the disease, the patient population, the dosing regimen required, the market competition, and the formulation and device possibilities. Finally, and sometimes above all, timelines and the need to be first to market may be important factors that guide a formulator on this difficult decision.

IX. SUMMARY AND CONCLUSIONS

The challenges for the successful delivery of protein pharmaceuticals by aerosol delivery to the lung is augmented by the need to couple a delivery device with the formulation. Thus, as in the case of liquid formulation, it is necessary not only to assure typical 2-year pharmaceutical shelf life, but also to prevent protein degradation due to the stresses that result from aerosolization. Development of dry powders adds complexity because of the need to have both biochemical and physical stabilty. The latter is required for ease of manufacture, filling into devices, and the requirement to attain a high fine particle fraction on delivery. This chapter has summarized some of the available devices and strategies for producing formulations that result in pharmaceutical shelf life and efficient delivery to the lung.

REFERENCES

1. R. T. Borchardt, series ed., In: T. J. Ahern and M. C. Manning, eds., In: Stability of Protein pharmaceuticals. Part A. Chemical and Physical Pathways of Protein Degradation. Pharmaceutical Biotechnology, Vol. 2. New York: Plenum press, 1992, p. 434.
2. J. L. Cleland, M. Powell, and S. J. Shire, The development of stable protein formulations: A close look at protein aggregation, deamidation and oxidation. Crit. Rev. Ther. Drug Carrier Sys. 10(4):307–377 (1994).
3. I. Gonda, Physiochemical principles in aerosol delivery. In: D. J. A. Crommelin, ed. Topics in Pharmaceutical Sciences 1991. Stuttgart: Medpharm Scientific, 1992, pp. 95–115.
4. P. R. Byron, Aerosol formulation, generation, and delivery using nonmetered systems. In: P. R. Byron, ed. Respiratory Drug Delivery. Boca Raton, FL: CRC Press, 1990, pp. 143–165.
5. G. Rudolf, et al., Mass deposition from inspired polydisperse aerosols. Ann. Occup. Hyg. 32:919–938 (1988).
6. J. Ranucci, Dynamic plume-particle size analysis using laser diffraction. Pharm. Technol. 16(10):109–114 (1992).
7. W. G. Gorman and F. A. Carroll, aerosol particle-size determination using holography. Pharm. Technol. 17(2):34–37 (1993).
8. P. D. Jager, G. A. De Stefano, and D. P. Mcnamara, Particle size measurement using right-angle light scattering. Pharm. Technol. 17(4):102–120 (1993).
9. J. A. Ranucci and C. F. Chen. Phase Doppler Anemometry: A technique for determining aerosol plume-particle size and velocity. Pharm. Technol. 17(6):62–74 (1993).
10. R. W. Niven, Aerodynamic particle size testing using a time-of-flight aerosol beam spectrometer. Pharm. Technol. 17(1):72–78 (1993).
11. A. R. Clark, The use of laser diffraction for the evaluation of the aerosol clouds generated by medical nebulizers. Int. J. Pharm. 115:69–78 (1995).

12. S. M. Milosovich, Particle size determination via cascade impaction. Pharm. Technol. 16(9):82–86 (1992).
13. P. J. Atkins, Aerodynamic particle size testing—Impinger methods. Pharm. Technol. 16(8):26–32 (1992).
14. USP/NF, Physical tests and determinations: Aerosols, Vol. USO XVIII. U.S. Pharmacopeia, Rockville, MD: United States Pharmacopeial Convention, Inc. 1992, pp. 3158–3178.
15. D. C. Cipolla and I. Gonda, Method for collection of nebulized proteins. In: J. L. Cleland and R. Langer, ed. Formulation and Delivery of Proteins and Peptides. Washington, DC: American Chemical Society, 1994, pp. 342–352.
16. R. W. Niven, J. P. Butler, and J. D. Brain, How air jet nebulizers may damage 'sensitive' drug formulations. In Abstracts, 11th Annual Meeting of the American Association for Aerosol Research, Oct. 12–16, 1992, San Francisco.
17. R. W. Niven, Delivery of biotherapeutics by inhalation aerosols. Pharm. Technol. 17(7):72–81 (1993).
18. D. C. Cipolla, et al., assessment of aerosol delivery systems for recombinant human deoxyribonuclease. S. T. P. Pharma Scie. 4(1):50–62 (1994).
19. S. P. Newman, et al., efficient delivery to the lungs of flunisolide aerosol from a new portable hand-held multidose nebulizer. J. Pharm. Sci. 85(9):960–964 (1996).
20. D. Cipolla, I. Gonda, and S. J. Shire, Characterization of aerosols of human recombinant deoxyribonuclease I (rhDNase) generated by jet nebulizers. Pharm. Res. 11:491–498 (1994).
21. J. Schuster, et al., The AERx aerosol delivery system. Pharm. Res. 14(3):354–357 (1997).
22. S. P. Newman, et al., The BINEB (final prototype): A novel hand-held multidose nebuliser evaluated by gamma scintigraphy. In: Annual Congress European Respiratory Society. Stockholm, Pharmaceutical Profiles Ltd, 1996.
23. T. E. Weston, A. W. King, and S. T. Dunne, Atomising devices and methods. In: World Intellectual Property Organization. Dunne Miller Weston Ltd., 1991. Patent WO 91/14468.
24. E. Ivri, Apparatus and methods for delivery of therapeutic liquids to the respiratory system. U.S. pataent 5,586,500 (1996).
25. J. M. Fine, et al., The role of titratable acidity in acid aerosol-induced bronchoconstriction. Am. Rev. Respir. Dis. 135:826–830 (1987).
26. J. R. Balmes, et al., Acidity potentiates bronchoconstriction induced by hypoosmolar aerosols. Am. Rev. Respir. Dis. 138:35–39 (1988).
27. R. Beasley, P. Rafferty, and S. T. Holgate, Adverse reactions to the non-drug constituents of nebulizer solutions. Br. J. Clin. Pharmacol. 25:283–287 (1988).
28. K. N. Desager, H. P. Van Bever, and W. J. Stevens, Osmolality and pH of anti-asthmatic drug solutions. Agents Actions. 31:225–228 (1990).
29. N. J. C. Snell, Adverse reactions to inhaled drugs. Resp. Med. 84:345–348 (1990).
30. G. Sant'Ambrogio, et al., response to laryngeal receptors to water solutions of different osmolality and ionic composition. Respir. Med. 85(suppl. A):57–60 (1991).

31. I. Gonda, et al., Characterization of hygroscopic inhalation aerosols. In: N. G. Stanley-Wood, ed. Particle Sixe Analysis. New York: Wiley Heyden, 1982, pp. 31–43.
32. I. Gonda and P. R. Phipps, Some consequences of instability of aqueous aerosols produced by jet and ultrasonic nebulizers. In: S. Masuda and K. Takahashi, eds. Aerosols. New York: Pergamon Press, 1991, pp. 227–230.
33. D. J. Godden, et al., Chemical specificity of coughing in man. Clin. Sci. 70:301–306 (1986).
34. B. Auffarth, et al., Citric ancid cough threshold and airway responsiveness in asthmatic patients and smokers with chronic airflow obstruction. Thorax, 46: 638–642 (1994).
35. D. Cipolla, et al., Formulation and aerosol delivery of recombinant deoxyribonu-cleic acid derived human deoxyribonuclease I. In: J. L. Cleland and R. Langer, eds. ACS Symposium Series 567, Formulation and Delivery of Proteins and Pep-tides. Washington, DC: Amercian Chemical Society, 1994, pp. 322–342.
36. C. Tanford, The Hydrophobic Effect. New York: Wiley, 1980.
37. J. Q. Oeswein, et al., aerosolization of Protein Pharmaceuticals. In: Procedings of the Second Respiratory Drug Delivery Symposium, University of Kentucky, 1991.
38. M. S. Hora, et al., Development of a lyophilized formulation of interleukin-2. Dev. Biol. Stand. 74:295–303 (discussion 303-6) (1992).
39. X. M. Lam, T. W. Patapoff, and T. H. Nguyen, The effect of benzyl alcohol on recombinant human interferon-gamma. Pharm. Res. 14(6):725–729 (1997).
40. R. Beasley, P. Rafferty, and S. Holgate, Benzalkonium chloride and bronchocon-striction [letter]. Lancet 2(8517):1227 (1986).
41. O. H. Berg, R. N. Henriksen, and S. K. Steinsvag, The effect of a benzalkonium chloride–containin nasal spray on human respiratory mucosa in vitro as a func-tion of concentration and time of action. Pharmacol. Toxicol. 76(4):245–249 (1995).
42. W. B. Klaustermeyer, F. C. Hale, and E. J. Prescott, Reproducibility of the response to diluent challenge in adult asthma. Ann. Allergy 43(2):84–87 (1979).
43. W. Wright, et al., Effect of inhaled preservatives on asthmatic subjects. I. Sodium metabisulfite. Am. Rev. Respir. Dis. 141(6):1400–1404 (1990).
44. Y. G. Zhang, et al., Effects of inhaled preservatives on asthmatic subjects. II. Benzalkonium chloride. Am. Rev. Respir.Dis. 141:1405–1408 (1990).
45. E. R. Stadtman, Metal ion-catalyzed oxidation of proteins: biochemical mecha-nism and biological consequences Free Radical Biol. Med. 9(4):315–325 (1990) [Erratum published in Free Radical Biol. Med. 10:3–48 (1991).
46. S. D. Anderson, et al., a novel bronchial provocation test (BPT) using a respirable dry powder of mannitol. In: Australian and New Zealand thoracic Society ASM, Perth Australia, 1996.
47. J. Y. H. Chang, T. Milby, and J. Q. Oeswein, Effects of metals and phenol on hGH degradation. Pharm. Res. 11(10):S81 (1994).

48. J. C. Lyda, FDA proposes sterility requirement of Inhalation solution drug products. PDA Let. November 1997, pp. 5–6.

49. M. B. Mecikalski and D. R. Williams, Dry powder inhaler. WO94/08552 (1994).

50. V. H. L. Lee, et al., Oral route of peptide and protein drug delivery. In: V. H. L. Lee, Peptide and Protein Drug Delivery. New York: Marcel Dekker, 1991, p. 891.

51. M. Eljamal, S. Nagarajan, and J. S. Patton, In situ and in vivo mehtods for pulmonary delivery. Pharm. Biotechnol. 8:361–374 (1996).

52. G. Rudolph, R. Kobirch, and W. Stahlhofen, Modeling and algebraic formation of regional aerosol deposition in man. J. Aerosol Sci. 21(supp. 1):S306–S406 (1990).

53. D. A. Edwards, et al., Large porous particles for pulmonary drug delivery. Science 276(5320):1868–1871 (1997).

54. H. K. Chan, et al., Spray dried powders and powder blends of recombinant human deoxyribonuclease (rhDNase) for aerosol delivery. Pharm. Res. 14:431 (1997).

55. J. Visser, Powder Technol. 58:1 (1989).

56. R. Niven, et al., Pulmonary delivery of powders and solutions of rhGCSF to the rabbit. Pharm. Res. 11:1101 (1994).

57. A. R. Clark, et al., The balance between biochemical and physical stabiltiy for inhalation protein powder: rhDNase as an example, In: R. Dalby and P. Byron, eds. Respiratory Drug Delivery V. Buffalo Grove, IL: Interpharm Press, 1996, p. 167.

58. J. F. Carpenter, et al., In: J. L. Cleland and R. Langer, ed. Formulation and Delivery of Proteins and Peptides. Washington, DC: American Chemical Society, 1993.

59. F. Franks and R. H. Hatly, Storage of materials. U.S. patents 5,098,893, (1992).

60. H. Levine, Biopharm. 5:36 (1992).

61. D. L. French, et al., Mositure induced state changes in spray-dried trehalsose/ protein formulations. Pharm. Res. 12(9):S-83 (1995).

62. C. Quan, In: Ninth Symposium of the Protein Society, Boston, 1995.

63. R. Wolf, The safety of inhaled therapeutic proteins. J. Aerosol Med. 11(4):197–219. 1998.

64. R. Platz, Improved process for preparing micronized polypeptide drugs. WO93/13752 (1993).

65. B. S. Chang and N. L. Fischer, Pharm. Res. 12:831 (1995).

66. A. R. Brown and J. G. Slusser, Propellant-driven aerosols of functional proteins as potential therapeutic agents in the respiratory tract. Immunopharmacology 28: 241–257 (1994).

67. A. R. Brown and J. A. Pickrell. Propellant-driven aerosols for delivery of proteins in the respiratory tract. J. Aerosol. Med. 8(1):43–57 (1995).

68. S. J. Shire, Stability characterization and formulation development of recombinant human deoxyribonuclease I[Pulmozyme®, (Dornase Alpha)] In: R. Pearlman and J. Wang, eds. Formulation, Characterization, and Stability of Protein Drugs. New York: Plenum Press, 1996.

69. A. L. Adjei and P. J. Carrigan, Pulmonary bioavailability of LH-RH analogs: Some biopharmaceutical guidelines. J. Biopharm. Sci. 3(1/2):247–254 (1992).

70. A. L. Adjei and J. Garren, Pulmonary delivery of peptide drugs: Effect of particle size on bioavailabiltiy of leuprolide acetate in healthy male volunteers. Pharm. Res. 7(6):565–569 (1990).

71. A. R. Clark, MDIs: The physics of aerosol formation. J. Aerosol. Med. 9(suppl. 1):S19–S25 (1996).

8

Formulation of Proteins for Incorporation into Drug Delivery Systems

OluFunmi L. Johnson
Alkermes, Inc.
Cambridge, Massachusetts

I. INTRODUCTION

Controlled delivery implies the incorporation of one or more elements of control on the release of an active ingredient from a dosage form to obtain well-defined pharmacokinetic profiles. A major advantage of controlled release formulations over conventional dosage forms is the ability to manipulate the components of the dosage form to obtain a particular absorption profile. Sustained release formulations are controlled release dosage forms that have been engineered to release the active ingredient over an extended period in a well-defined and reproducible fashion (1). A wide variety of dosage forms have been developed for the delivery of conventional small-molecule drugs via the oral, parenteral, buccal, transdermal, ocular, intravaginal, intrauterine, and nasal routes. Each of these routes of delivery has its own unique advantages and challenges, which are a function of the physiology of the sites. The delivery of proteins via any of these routes requires special considerations that must be borne in mind in the development of protein formulations. This chapter focuses on sustained delivery and the unique challenges of formulating therapeutic proteins and peptides in sustained release dosage forms.

The production of proteins by recombinant technology was an important milestone in the development of a new class of therapeutic agents. The availability of these new biopharmaceuticals has made their formulation and delivery an important part in the treatment of various disease states. For example, a protein drug with a very short half-life may require frequent injections for efficacy. Often, however, frequent injection of high doses is not practical, and the fluctuating serum drug levels may cause unacceptable side effects. The short-half-life may be dealt with by chemically modifying the protein, by coupling the protein to a moiety such as polyethylene glycol or by developing a sustained release formulation that would release therapeutic concentrations of the drug over an extended period.

Proteins differ from conventional small-molecule drugs in several respects, but one of the most important differences is the complexity of protein structure. The body has evolved ways of maintaining the specificity of action of proteins, and this is controlled to a large extent by the different levels of organization (viz., primary, secondary, tertiary, and quaternary structures). Because of the close correlation of protein efficacy with the molecular three-dimensional structure, it is essential to maintain structural integrity through all the formulation steps until the drug is released from the dosage form at the site of delivery. Otherwise the activity of the protein or peptide drug is lost.

Because of the fragile nature of proteins and peptides, the processes involved in the fabrication of the drug delivery system may damage the protein and, therefore, reduce its biological activity or render the protein immunogenic. For example, aggregated human growth hormone has less biological activity than the native monomeric form (2). This chapter deals with some of the fabrication methods, the potential for affecting protein integrity, and approaches to overcome the problems that arise. Formulating proteins and peptides for conventional dosage forms such as powders or solutions is a difficult enough undertaking; but formulating protein drugs for controlled release formulations entails additional hurdles, and these must be considered during the formulation development process.

The problems associated with assuring the stability of a protein or peptide drug in various matrices have been discussed (3,4). First, the protein must survive the processes involved in the fabrication of the delivery matrix. This is followed by an extended period within the delivery matrix, in situ, while the entrapped drug is released. During this period in vivo, the protein within the matrix is subjected to potentially unfavorable conditions (of pH, protein concentration, etc.) as degradation of the matrix proceeds. Therefore, it is not enough to have a stable injectable protein solution, because additional or completely different stabilization strategies are often required for successful formulation into a controlled release formulation.

Development of a formulation is further complicated by the effect of the site of delivery, so a sustained release formulation for nasal delivery may not be appropriate for delivery by subcutaneous injection. There is usually a tissue response to the implantation of a delivery matrix (even if it is biocompatible, as with polylactic and polyglycolic polymers) because it is perceived as a foreign material. The concentration of phagocytic cells in response to the delivery device and the proteolytic enzymes they produce may destroy the encapsulated protein as it is released, possibly resulting in a reduction in the bioavailability of the protein drug compared to a bolus injection.

This chapter provides an overview of protein degradation pathways and discusses protein stability issues associated with the fabrication and delivery of protein drugs in sustained release systems. Armed with this information, the formulator may then adopt a rational and informed empirical approach. However, in spite of a well-thought-out strategy, some protein drugs simply are not amenable to delivery via sustained release dosage forms.

II. PROTEIN DEGRADATION PATHWAYS

The degradation mechanisms proteins undergo can be divided into two classes, chemical and physical. In chemical degradation, the native structure of the protein is changed by modifications to the primary structure. In physical degradation, the native structure of the protein is modified by changes to the higher order structure of the protein (secondary, tertiary, or quaternary structure). The degradation may be brought about by aggregation, adsorption, unfolding, or precipitation. Chemical degradation usually involves bond cleavage, and the product is a new chemical entity. Chemical degradation processes include deamidation, oxidation, racemization, disulfide exchange, and hydrolysis and are usually preceded by a physical process such as unfolding, which then makes usually inaccessible residues available for chemical reactions.

A. Physical Instability

Proteins, because of their polymeric nature and the ability to form higher order structures, can undergo non-chemical changes (i.e., changes that do not affect the primary structure), which can alter their biological activity. The primary structure of a protein determines the native secondary and tertiary (and higher order) conformation. In general, in globular proteins, hydrophobic residues are buried in the interior and hydrophilic residues are available for interaction with the aqueous solvent. Denaturation refers to the loss of this globular structure and leads to protein unfolding, the extent of which may or may not result in the loss of secondary structure. Once unfolded, the protein may adsorb to surfaces expose amino acid residues that normally would be protected to chemical modification, or the protein may aggregate by interaction with other protein molecules.

Denaturation may be reversible or irreversible and is caused by changes in the environment of the protein, such as in temperature, or pH changes, the introduction of interfaces by the addition of organic solvents, or the introduction of hydrophobic surfaces. In reversible denaturation, the protein refolds once the denaturing stimulus has been removed, (e.g., unfolding when the temperature is increased and refolding correctly once the temperature has been reduced). In irreversible denaturation, the native conformation is not regained by the removal of the stimulus, although some activity may be regained by controlling the pathway by which the protein is returned to the nondialyzing state (e.g., addition of a denaturant such as urea followed by dialysis) (5).

Precipitation is the final result of self-association or aggregation of protein molecules. The aggregation of insulin has been well characterized and is thought to depend on unfolding of the insulin molecule (6).

B. Chemical Instability

Chemical degradation processes may occur at several points during the formulation and delivery of an encapsulated protein drug. It is important to be aware that the manifestation of a degradative process may occur after the triggering step. For example, the pH of the buffer before freeze-drying can affect the stability of the lyophilized protein formulation (7).

1. Oxidation

Tryptophan, methionine, cysteine, histidine, and tyrosine amino acid side chains contain functionalities that are susceptible to oxidation. Methionine and cysteine can be oxidized by atmospheric oxygen and fluorescent light (2), and oxidation has been observed both in solution (8) and in the solid state (9). Oxidation of the methionine residues may cause a loss of bioactivity and, in the case of cysteine residues, formation of non-native disulfide bonds. Oxidation by atmospheric oxygen or autoxidation can be accelerated in the presence of certain metal ions such as copper and iron. Methionine residues under acidic conditions are especially prone to oxidation by reagents such as hydrogen peroxide, producing methionine sulfoxide. Oxidation by peroxide may be an issue when hydrogen peroxide is used to sterilize formulation vessels or the formulation area (10). Under conditions of higher pH, other groups such as disulfide and phenol groups may undergo oxidation reactions.

2. Deamidation

The hydrolysis of a side chain amide on glutamine and asparagine residues to yield a carboxylic acid is called deamidation. This reaction has been extensively studied and is widely observed in therapeutic proteins and peptides (5). Some conditions, such as increase in the temperature and pH a protein is likely to experience during formulation, have been shown to facilitate deamidation. The deamidation process is important because of the potential loss in protein activity or function. Deamidation contributed to the reduction in catalytic activity of lysozyme (5) and ribonuclease at high temperatures (11).

3. Peptide Bond Hydrolysis

Aspartic acid residues have been implicated in the cleavage of peptide bonds, which in turn has led to a decrease in biological activity. When lysozyme was heated to 90–100°C, at pH 4, the loss in biological activity was attributed to hydrolysis of Asp-X bonds (5).

4. Disulphide Exchange

Many therapeutic proteins contain cysteine residues that form disulfide bonds. These bonds are important components of the structural integrity of proteins (12). Incorrect linkages of these disulfide bonds often lead to a change in the three-dimensional structure of the protein and therefore its biological activity. The reaction proceeds in both acidic and alkaline media, but the mechanisms are different. In neutral and alkaline media, the reaction is catalyzed by thiols. Thiols may be introduced during formulation (e.g., mercaptoethanol as an antioxidant) or by degradation of existing disulfide bonds via β-elimination of cysteine residues (13). The aggregation of lyophilized formulations of bovine serum albumin, ovalbumin, β-lactoglobulin, and glucose oxidase was attributed to disulfide interchange (14).

III. INCORPORATION INTO THE DRUG DELIVERY MATRIX

The fabrication of drug delivery devices or dosage forms often involves steps or harsh procedures that may entail temperature elevation, pH changes, and the generation of conditions of high shear (15). These processes may denature the protein with a subsequent loss in bioactivity. Considering drug encapsulation into microparticulates, the incorporation process may involve the formation of emulsions and the associated interfaces and potential for denaturation, the use of mechanical agitation to facilitate formation of the droplets, exposure to organic solvents, and significant fluctuations in temperature, concentration, and pH. There has not been a systematic study of the effects of incorporation processes on protein denaturation or degradation. There are, however, reports on the damaging effects of emulsification and organic solvents on proteins (16,17), and in other cases, proteins seem to retain most of their biological activity after incorporation (18,19).

A. Emulsification

The deleterious effects of emulsification have been addressed by several authors (16,17). The protein or peptide drug is often incorporated into a matrix that is insoluble in water, but since the protein or peptide drug is usually more water-soluble, two solvents are employed, one for the matrix material and another for the protein drug. Briefly, the water soluble drug is dissolved in an aqueous solution (or water) and the polymer is dissolved in an organic solvent such as ethyl acetate or methylene chloride. The two solutions are mixed in the appropriate ratio to create a water-in-oil emulsion. This primary emulsion is then emulsified into an aqueous solution containing an emulsifier with a hydrophile–lipophile balance (HLB) number greater than 10. The final product is a water-in-oil-in-water emulsion (w/o/w) (20). There are variations on this basic approach that use a range of aqueous and organic solvents and a range of aqueous phase emulsifiers. The organic solvent is then removed from the emulsion by evaporation under reduced pressure, filtration, or a moderate increase in temperature. The microspheres are then harvested and dried. The formation of the microparticle dosage form is a complex process, and the protein is subjected to potentially destabilizing conditions, from the intense mechanical stress of emulsification to exposure to hydrophobic interfaces in the emulsion and the solvent removal process.

Emulsification is achieved by disrupting a mixture of the aqueous solution that contains the protein drug with other water-soluble excipients and the organic water-immiscible phase that contains the polymer or matrix material. Emulsion droplets are formed by the input of energy, which may be mechanical (as with a homogenizer) or ultrasonic (with the use of an ultrasonic probe). The application of the energy source is accompanied by one or more processes including high shear, cavitation, and high temperature. In addition, the formation of the emulsion droplets results in the formation of a large interfacial surface at which the protein can accumulate, unfold, and denature. Formation of emulsion droplets is a violent process and can be initiated by film formation, generation of surface waves, deformation of the aqueous–organic solvent interface, and cavitation. The turbulence eddies that are set up during high speed homogenization cause disruption of the interface and droplet formation. Cavitation is the main process by which ultrasonic waves form emulsion droplets, and it is the sudden formation and collapse of bubbles containing vapor that results in the generation of high pressures, on the order of 10^{10} Pa and high intensity shock waves. The combination of high pressures, intense shock waves, and the dissipation of both over a short time period causes the formation of droplets. The energy that is generated is accompanied by a significant increase in temperature. Although these extreme conditions may be short-lived, nevertheless, significant damage may be done to the protein, and loss of protein activity or potency is often observed after encapsulation (21). The high shear rates that are generated during any emulsification process may be enough to cause foaming [adsorption of the protein at the air–water interface (9) or at the aqueous–non aqueous interface of the emulsion droplets]. In either case the result may be aggregation and eventually precipitation.

B. Coacervation

Microparticles containing a protein may be formed by coacervation. In simple coacervation, a hydrocolloid containing the protein is desolvated by the addition of another substance (such as a salt or an alcohol), which competes for the solvent by virtue of its higher hydrophilicity or concentration. In complex coacervation, the charge of the hydrocolloid is opposite to that of the competing substance, so that when the latter is added, a complex of the two is formed and the mixed coacervate is separated by dilution (22). The microcapsules formed are then "cured" by the addition of a cross-linking agent such as glutaraldehyde. Hydrocolloids that are used in pharmaceutical formulations include gelatin, acacia, and chondroitin sulfate (23). For example, cytokines have been encapsulated in gelatin–chondroitin sulfate microspheres. The cytokine IL-2 (interleukin 2) was dissolved in chondroitin sulfate solution and a solution of gelatin was added to form microspheres. The microspheres were then cross-linked by glutaraldehyde. In addition to the stresses the protein molecules undergo as a result of pH changes during coacervation, the cross-linking step with glutaraldehyde is indiscriminate and cross-links the microsphere matrix as well as the encapsulated protein or peptide drug. This lack of selectivity accounts for some of the loss in bioactivity of protein drug encapsulated in microspheres that utilize chemical cross-linking.

C. Extrusion and Spraying Methods

Extrusion or spraying methods may be used in the fabrication of drug delivery systems to form droplets or monolithic injectable delivery devices (15). In the former case, when extrusion or spraying is employed to form microspheres, the core material or matrix containing the protein drug, incorporated as a solution or particulate, is ejected from the orifice of a fine tube, syringe, or nozzle to form microdroplets. The size of the droplets and, therefore, the final dosage form, depends on the properties of the liquid (melt, solution, or suspension) to be sprayed and on the operating conditions of the extruder, such as orifice diameter and jet velocities (24). The main considerations as far as the stability or integrity of protein is concerned are the processing conditions such as the melting temperature if a melt extrusion method (15) is employed and the effect of the high shear forces that may be generated from the orifice at high jet velocities.

A method using a cryogenic process to produce microparticles containing proteins has been described (25). In this method, the protein drug is incorporated as a lyophilized powder and all manipulations involving the matrix polymer (polylactide co-glycolide: (PLG) and the protein are performed at low temperatures ($\leq -80°C$). This approach has two important benefits. First, the protein is in the more stable dry form, and any degradation processes it is liable to undergo even in the dry state are hindered by the very cold temperatures. Second, because there are no aqueous phases, the protein in not subjected to freeze/thaw stresses. Provided a stable lyophilized formulation is available, this process effectively solves the issue of protein denaturation during fabrication.

D. Polymerization

Hydrogels are polymeric delivery systems that swell when they come into contact with water or aqueous solvents. The extent of swelling is determined by the cross-linking density of the polymer network, the polymerization conditions, and the polymer–solvent interaction parameter. Hydrogel delivery systems may be prepared by mixing the monomer with the drug, an initiator, and a cross-linking agent and allowing the monomer to polymerize (26). The polymerization and cross-linking steps may be a problem because agents used (e.g., γ-radiation) may have deleterious effect on the integrity of the protein.

Gel drug delivery systems containing proteins and peptides have been fabricated by preparing a formulation of the peptide or protein drug with the monomer and inducing polymerization by the application of electromagnetic radiation. An example is the intravascular delivery of proteins via hydrogel films that were photo-polymerized in situ on the inner surface of blood vessels. The hydrogel precursor consisted of a central polyethylene glycol chain with lactic acid units on either end and capped with a reactive acrylate group. The protein drug was dissolved in the aqueous precursor solution and applied to the surface of the pretreated blood vessel. The solution was then polymerized in situ by irradiation from a xenon arc lamp (400–600 nm, 35 mW/cm^2, 10 s) (27).

IV. PROTEIN STABILITY WITHIN THE DELIVERY MATRIX: INTERNAL AND EXTERNAL FACTORS

In addition to the considerations given to protein stabilization during the incorporation process, the environment the protein encounters within the delivery device and the site of delivery in the body should be factored into the formulation development. For sustained release or depot formulations, the objective is to have a reservoir of the protein or peptide drug that will be released over a period of days to months. These potential interactions should be taken into account during protein formulation. Parenteral introduction of a delivery device or dosage form will induce a tissue response. It was shown (28) that PLG microcapsules injected intramuscularly were completely engulfed in a thin layer of connective tissue, and there was evidence of infiltration by macrophages. It is feasible that the influx of these macrophages may cause degradation of the encapsulated protein and protein released in the vicinity of the device. It has been suggested that these macrophages are capable of producing proteolytic enzymes (29), which may result in the release and circulation of altered, inactive, or immunogenic forms of the encapsulated protein or peptide. Systematic studies on the effects of tissue response on the bioavailability of incorporated proteins have not appeared in the literature.

There are also issues related to the physiology and the normal function of the delivery site. For example, to obtain appreciable bioavailability after nasal administration, penetration enhancers are often coadministered with protein and peptide drugs (30). These enhancers may also interact with the protein, and the potential effect of the stability or activity of the protein should be evaluated at the formulation development stage.

A. Interactions Between the Delivery Matrix and the Protein

The incorporation of protein pharmaceuticals into solid delivery matrices exposes them to a high surface-to-volume environment, and there is ample opportunity for adsorption to the delivery device. One obvious drawback of this type of adsorption is that it may severely limit the amount of free unadsorbed protein that is available for release. It has been shown that salmon calcitonin adsorbs to the surface of PLG matrices (31). Another consequence of adsorption may be surface-induced changes in the three-dimensional structure of the protein that could result in loss of biological activity. A number of factors affect the adsorption of proteins to the solid interface. These include the charges on both the protein and the surface, effects of the environment (e.g., pH, ions present, specificity of adsorption, surface area, temperature) (32). The surface activity of a protein depends on its amino acid composition. This primary structure determines which amino acid residues are exposed to the surface in the three-dimensional structure and, therefore, available for interaction with the delivery matrix. Since proteins are ampholytes, the pH and ionic strength of the surrounding medium will determine the net surface charge and, therefore, the nature of the interaction with the surface. Consequently, there will be a stronger interaction between the matrix surface and the protein when both possess opposite charges.

In aqueous solution, the three-demensional structure of a protein in its native conformation results in the burial of hydrophobic residues in the interior and the exposure to the aqueous solution of more hydrophilic amino acid residues. However, when the same protein comes into contact with a hydrophobic surface, a driving force will be exerted on the hydrophobic residues that are normally buried within the three-dimensional structure to interact with the surface, causing unfolding or other structural rearrangements. Both the extent and effect of adsorption on the biological activity of a protein are protein specific, and the adsorption of antibodies on solid matrices in immunoassays suggests that interactions with solid matrices does not automatically result in loss of activity. The adsorption isotherms of proteins to solid surfaces displays a saturation phenomenon. There is an initial rapid phase of adsorption that eventually reaches a plateau at higher protein concentration at a given temperature. The saturation point approximates a monolayer, suggesting that the number of adsorption sites is fixed.

One of the most effective ways of formulating a protein to reduce adsorption to the delivery matrix or surface is by incorporation of a surface active agent or the addition of another protein to compete for adsorption sites.

Addition of albumin to insulin solution was found to reduce adsorption of the latter to solid surfaces. Addition of surfactants such as polysorbate or sodium dodecyl sulfate reduces the interfacial tension at the solid–liquid interface and therefore the driving force for the protein to adsorb (33).

B. Internal Environment of the Delivery Matrix

Many sustained delivery system matrices utilize biodegradable materials; therefore, the protein may be exposed to different environments as the delivery matrix degrades over time. For example, a number of implantable devices have been fabricated from polyesters such as polylactic and polyglycolic acid, which degrade to lactic acid and glycolic acid. It is feasible that the generation of these degradants could cause an increase in the acidity of the interior of the microsphere. There have been few studies to investigate the internal pH of these devices (34,35) but once implanted, it is unlikely that the devices have their interiors totally isolated from the perfusion and buffering capacity of physiological fluids. The degree of isolation between the interior of the device and its exterior will depend on the porosity and geometry of the device. It seems intuitive that if a large molecule such as a protein is able to exit the matrix (by diffusion through polymer networks or pores), then small molecules such as buffer salts or the soluble monomeric or oligomeric degradation products of biodegradable matrices can diffuse in and out of the matrix. If however, release from the matrix depends exclusively on polymer matrix erosion, such a device may not be suitable for protein delivery because of the potential buildup of high levels of degradation products and pH extremes. It may be possible to incorporate a buffering system into the matrix. For example, excipients may be incorporated into a PLG delivery matrix to counteract the increase in acidity produced by the degradation of the polymer into lactic and glycolic acids.

C. External Factors: Site of Delivery

The site of delivery of the dosage form should be considered early in the formulation of the protein, since the characteristics of the route of delivery may profoundly affect the efficiency of delivery of the biopharmaceutical in active form and in therapeutically relevant concentrations. The presence of a parenteral implant induces the migration of macrophages and other white blood cells to the site of injection or implantation. In the case of PLG microspheres, the microspheres were completely surrounded by a thin layer of cells within a few days of injection (28). It has also been suggested that the presence of this layer of cells may be a determinant of release of encapsulated drug (36). Such a response may affect the degradation kinetics of the matrix or the amount of protein degraded at the site of injection, hence the bioavailability of the protein or peptide. The delivery of protein and peptide biopharmaceuticals across mucosal membranes (e.g., nasal and vaginal) almost always requires the incorporation or coadministration of excipients to increase absorption and bioavailability.

1. Transmucosal Delivery

Transmucosal delivery of proteins and peptides refers to the release of the protein or peptide drug incorporated in a delivery matrix across a biological membrane to achive therapeutic serum levels.

2. Nasal Delivery

There are three main challenges to controlled protein or peptide delivery; these are overcoming the nasal clearance mechanism, the absorption barrier and enzymatic activity. These factors should all be taken into consideration during the development of a dosage form for nasal delivery. The nasal cavity is a promising site for the delivery of drugs because of the large surface area of the cavity and because there is easy access to the systemic circulation while the hepatic circulatory first-pass effect is avoided (37). However, the very physiology of the nasal cavity poses several challenges for protein and peptide delivery. First, the nasal epithelium is covered by cilia and a mucosal layer, which together form the mucociliary clearance system. The function of this system is to rapidly remove any foreign material, including gel, liquid, and solid dosage forms, before absorption has a chance to occur. Assuming that the mucociliary clearance hurdle is overcome (e.g., by means of a bioadhesive delivery matrix) (38), there is an absorption barrier to large molecules, especially proteins, and peptides to a lesser extent. This problem is dealt with by the addition to the protein formulation of penetration enhancers. There is significant proteolytic activity in the nasal cavity, which can be overcome by the use of protease inhibitors (39).

3. Oral Delivery

There are two main considerations in the development of protein or peptide formulations for oral delivery, namely, the enzymatic and absorption barriers. One of the major functions of the digestive system is to break down complex molecules such as proteins and carbohydrates into amino acids and sugars, which are easily absorbed. The same mechanism makes effective delivery and absorption of intact, bioactive protein drugs difficult. Hydrolysis of protein drugs occurs in several parts of the intestinal tract, starting in the stomach with its secretion of gastric juices all the way into the small intestine. Oral delivery of smaller peptides has been more successful because proteases (pancreatic) are less active against them. Several peptides are orally active, such as cyclosporine and vasopressin analogs (40). These are peptides that have about 10 amino acid residues. Cyclosporine is a very hydrophobic peptide and is administered orally, in an oily vehicle, as an immunosuppressant for transplant patients. Apart from these peptide drugs, there have been few reports of successful delivery of proteins via the oral route because of the enzymatic barrier. A new class of molecules that alter the conformation of proteins reversibly and facilitate their transport across intestinal mucosa was described by Leone-Bay et al. (41). These researchers demonstrated the absorption of human growth hormone after oral administration with these N-acylated, non-α, aromatic amino acid compounds.

4. Parenteral Delivery

Most sustained release formulations of proteins are administered via a parenteral route: by subcutaneous (Lupron Depot) or intramuscular injection, or by implantation. The delivery device is perceived as a foreign body and elicits a tissue response even if the matrix material is biocompatible, as is the case with poly(lactide-glycolide) implants.

V. GENERAL PROTEIN FORMULATION STRATEGIES

The preceding paragraphs outlined some of the important processes that cause protein instability and subsequent loss or reduction in biological activity, and the stresses to which a protein drug is likely to be subjected during incorporation and incubation in the delivery matrix at the site of delivery. In this section, an attempt will be made to identify some general approaches that may be useful in developing sustained release formulations of proteins. It should be borne in mind, however, that the development of a particular protein formulation still must be determined empirically, with consideration of the final configuration, because of the potential interactions of the protein with the delivery device and the influence of the site of administration. For example, a stable formulation of a protein drug for weekly delivery in a retrievable silicone device into the subcutaneous space may not be directly transferable into a once-a-month biodegradable device administered intramuscularly. This is because of the potential effects of protein dose, interactions between the protein

and the device matrix, the effect of the changing environment as the biodegradable matrix erodes, and the difference between the subcutaneous and intramuscular biological environments. There is a large body of literature on protein-stabilizing excipients, and there are a number of excellent reviews on the
subject (12,42), and these are a useful starting point.

Stabilizing additives used in the formulation of proteins are diverse and
include proteins, sugars, polyols, amino acids, chelating agents, and inorganic
salts. These additives can stabilize the protein in solution and also in the frozen
and dried states, although not all the additives confer stability on the protein
or peptide under all three conditions. The stabilization mechanism in the solution of frozen state is different from that which occurs in the dried state, and
carbohydrates in particular have the ability to stabilize in the dried state (43).
This differential ability to stabilize the protein in various states indicates that
if a protein drug is to be transformed from a liquid to a powder by lyophilization or spray-drying in the course of incorporation into the delivery system,
more than one additive may be required to stabilize the protein. A stable liquid
protein formulaton is a prerequisite because it will usually form the starting
bulk material before any other manipulations are undertaken, whether subsequent steps involve lyophilization or emulsification. Stability in this state does
not imply long-term or storage stability, only stability that lasts long enough
for the protein to be handled and transformed into the next step.

Many additives used in protein formulation [e.g., sugars, salts, amino
acids (44) and polyols (45)] stabilize proteins by a similar mechanism (46).
Sugars such as trehalose, sucrose, maltose, and glucose have been used as
protein stabilizers and their addition causes an increase in the transition temperature of a number of proteins such as collagen, ribonuclease, and ovalbumin
(47). The diversity of proteins stabilized by sugars suggests that the stabilizing
effects of sugars may not be specific for particular proteins. Similarly, salts
such as potassium phosphate, sodium citrate, and ammonium sulfate have also
been shown to increase the transition temperature of proteins and therefore to
stabilize the protein. All these additives increase the self-association of protein
molecules and reduce the solubility of the protein. It has been suggested that
the mechanism that is common to all these cosolvent stabilizers is the exclusion of the additives from the surface of the protein (47). On the other hand,
stabilizers such as magnesium chloride bind to the protein surface. Cyclodextrins have also been used as stabilizing excipients in protein formulations.
Hydroxypropyl cyclodextrin (HPCD) stabilized formulations of porcine
growth hormone against thermal and interfacial denaturation (48). The mechanism is poorly understood but is thought to involve changes in the properties
of the solvent. The addition of heparin, a polyanion, stabilized acidic fibroblast

growth factor by increasing the unfolding temperature by 15–30°C by a direct interaction between the polyanion and the protein (49).

Surfactants are frequently added as stabilizers to protein formulations, and they serve various useful functions in sustained release formulations. Several commercial preparations of proteins (e.g., Nutropin® a recombinant human growth hormone from Genentech, Inc.) contain surfactants such as the polysorbates (2). Because proteins are amphipathic, they tend to adsorb and accumulate at interfaces. This interfacial adsorption may lead to unfolding and eventually to loss of solubility, aggregation, and biological activity (50). The addition of surfactants may be an important step in stabilizing the three-dimensional structure of a protein during incorporation into a delivery device—for example, during emulsification processes (19). The utility of the surfactant polysorbate 20 in stabilizing human growth hormone incorporated in a PLG matrix was demonstrated by Cleland and Jones (51).

Sustained delivery devices may have relatively hydrophobic surfaces, which may induce unfolding of the protein by adsorption and exposure of the more hydrophobic residues, which normally would be embedded in the interior of the three-dimensional structure of the protein. In this situation, addition of a surfactant with a high HLB would reduce protein adsorption because the surfactant itself would adsorb to the device surface via its hydrophobic moieties, leaving the hydrophilic moieties exposed to the surrounding milieu (33). The ability of a surfactant to stabilize a protein upon rehydration has also been demonstrated (52). Therefore, the addition of a surfactant can stabilize a protein against denaturation during several stages from incorporation to release at the site of delivery.

The addition of certain metals has been shown to confer stability on proteins (53,58). Cunningham et al. showed that certain transition metals stabilized human growth hormone (hGH) against urea induced denaturation. They hypothesized that the presence of zinc in the secretory granules of the anterior pituitary (where growth hormone is secreted) stabilize the hormone during storage before it is released into the circulation. Johnson et al. (54) showed that an hGH-Zn complex was a viable formulation for encapsulation into a sustained release formulation of polylactide co-glycolide microspheres. The hGH-Zn complex was more stable in PLG microspheres that hGH in the uncomplexed form (55). The authors demonstrated in primates (rhesus monkeys) that the protein released over a one-month period was active as measured by a serum hGH and serum levels of insulin-like growth factor I (IGF-I), a pharmacodynamic marker for hGH.

While the addition of metals may stabilize some proteins, the presence of metals may catalyze the oxidation of cysteine residues. To overcome this, chelating agents such as EDTA were added to stabilize a formulation of acidic fibroblast growth factor (49). Under accelerated storage conditions at 45°C, the addition of EDTA stabilized a freeze-dried formulation of ribonuclease A (56). Where there is evidence that the presence of trace amount of metals accelerates protein degradation, the addition of chelating agents should be considered because protein incorporation methods such as ultrasonication may release trace amounts of metals from the ultrasonic device.

Changing the state of a protein or peptide from the liquid to the solid state by lyophilization or spray-drying increases the storage stability of proteins (42). In addition to this obvious advantage, incorporation of the protein into a delivery device in the form of a lyophilized powder may also significantly reduce the potentially damaging stresses to which a protein solution would be subjected. For example, the protein in the solid state would be less susceptible to shear forces that occur during an emulsification procedure or denaturation at oil–water interfaces because the protein would not be in solution. However, special precautions should be taken during freeze-drying because the drying process itself will expose the protein to destabilizing stresses; therefore, suitable excipients should be included in the formulation (57).

Additives that are normally added to stabilize a formulation that will be lyophilized may include a bulking agent (sucrose, dextrose, mannitol) to prevent collapse of the freeze-dried cake. Cryoprotectants are also added to lyophilized formulations to stabilize the protein to freezing (51). Some of the salts, polyhydric alcohols, and sugars (described earlier) stabilize a protein to withstand the effects of freezing but additional stabilizers may be required during the drying and rehydration steps. Carbohydrates, disaccharides in particular, stabilize proteins during drying (43). Infrared spectroscopy measurements showed that hydrogen bonding occurs between these disaccharides and the proteins (bovine serum albumin, lysozyme). As the protein becomes dehydrated during lyophilization, these disaccharides are thought to act as water substitutes and are able to hydrogen bond to the proteins, thereby maintaining the integrity of the protein. The physical state of the carbohydrate stabilizer is also thought to be important, since the amorphous state is more effective than the crystalline state in stabilizing the protein (9,58). In the amorphous state, the protein has less mobility and chemical reactions are, therefore, significantly reduced. A surfactant is often added to facilitate dispersion of the protein and to reduce adsorption (and unfolding) of the protein to the walls of the dissolution vessel. The advantages of including a surfactant have already been discussed.

It should be borne in mind that a lyophilized formulation is not stable indefinitely; over time the protein may become denatured and lose activity.

The effect of excipients on moisture-induced aggregation of human serum albumin under conditions designed to simulate the environment inside sustained delivery devices has been investigated (7,59,60). The authors showed that the aggregation can be induced via both covalent and noncovalent routes and suggested some rational approaches based on an understanding of the underlying aggregation mechanisms. These approaches involved the effect of the buffer pH before lyophilization, the addition of high molecular weight polymers such as dextran and carboxymethylcellulose, the modification of the hydrophilic properties of the delivery matrix to reduce water uptake, and the manipulation of the protein molecule itself to chemically alter the residues involved in the degradation pathway. The studies described underline the im-

Table 1 General Formulation Strategy for the Incorporation of a Hypothetical Protein into a Delivery Matrix[a]

Protein attribute	Formulation approach
Unstable in solution	Consider lyophilization using cryoprotectants and incorporating drug into delivery matrix as a solid powder.
Adsorbs to delivery matrix (PLG)	Consider the incorporation of hydrophilic surfactants (polysorbate 20, polysorbate 80, Pluronic F.68®).
High protein concentration required in delivery system-prone to aggregation	Consider adding surfactant to reduce self-association.
	Consider incorporating as a less soluble species (complexation with metal: e.g., zinc). *Note*: Reduced solubility species must be completely reversible to the starting material without loss of activity.
Poor stability at low pH	Lyophilize or formulate in high pH buffer; incorporate soluble basic salt in delivery matrix to neutralize acid degradation products of delivery matrix.
	Consider using microporous delivery matrix rather than monolithic device.
Heat sensitivity	Consider low temperature homogenization encapsulation process.

[a] Background information: polylactide-*co*-glycolide delivery matrix, parenteral route of delivery—subcutaneous or intramuscular; 3D structure of protein required for biological activity, protein solubility is >20 mg/mL.

portance of understanding the underlying degradation processes and the more information there is on these degradation pathways, the more likely it is that a rational approach can be successfully applied.

The site of delivery should be considered during the formulation of protein and peptide drugs. It may be necessary to incorporate excipients such as protease inhibitors and absorption enhancers (61). These excipients increase the bioavailability of peptides and proteases when delivered to a mucosal surface such as the nasal cavity, where normal physiological mechanisms render the absorption of protein (and peptide to a lesser extent) inefficient (62). Absorption enhancers such as fatty acids and bile salts and enzyme inhibitors such as aprotinin and soybean trypsin inhibitor have been used (61). Since, fatty acids tend to undergo autoxidation reactions, with the generation of reactive oxygen species, the potential for the degradation products of excipients to affect the stability of the protein or peptide drug should be carefully monitored. Cyclodextrins, which are used to stabilize proteins, have also been used to enhance absorption of peptides across the nasal mucosa (63). It was suggested that the formation of an inclusion complex with leucine enkephalin protected the peptide against enzymatic degradation and enhanced absorption.

The principles described in this section may be used to develop a general strategy for incorporating a protein in a delivery system. Consider a hypothetical protein for nasal delivery in a PLG matrix. More likely than not, unforeseen issues will arise and require to be dealt with. Table 1 outlines a general approach to preliminary formulation development.

VI. CONCLUSIONS

The successful formulation of protein and peptide drugs for drug delivery systems should include considerations of the processes and environments the drug will encounter during incorporation into the delivery matrix or device, the internal environment of the device before release at the site of delivery, and interactions between the delivery device and the biological response to the delivery device. There is no specific approach to ensure the successful formulation of proteins for sustained delivery. Rather, there should be a thorough understanding of the degradation processes to which a particular protein drug is prone and the possible stresses to which a particular incorporation process, delivery matrix, and site or route of delivery will expose the protein. In general, protein incorporation processes that involve interfaces, as in emulsions, should be avoided. If the protein can be incorporated in the solid form, as a lyophilized powder, the chances of degradation are lessened. And if, as in the case of human growth hormone, the protein can be encapsulated in a stable form with reduced aqueous solubiltiy, the potential for degradation while awaiting release at the site of delivery is greatly reduced.

ACKNOWLEDGMENTS

I am grateful to Dr. Stephen Zale for helpful discussions during the preparation of the manuscript.

REFERENCES

1. R. Langer, Polymer-controlled drug delivery systems. Acc. Chem. Res. 25(10): 537–542 (1993).
2. R. Pearlman and T. A. Bewley, Stability and characterization of human growth hormone. In: Y. J. Wang and R. Pearlman eds. Stabiltiy and Characterization of Protein and Peptide Drugs: Case Histories. New York: Plenum Press, 1993, pp. 1–57.
3. C. G. Pitt, The controlled parenteral delivery of polypeptides and proteins. Int. J. Pharm. 59:173–196 (1990).
4. W. Lu and T. G. Park, Protein release from poly(lactic-co-glycolic acid) microspheres. J. Pharm. Sci. Technol. 49(1): 13–19 (1995).
5. M. C. Manning, K. Patel, and R. Borchardt, Stability of protein pharmaceuticals. Pharm. Res. 6(11):903–918 (1989).
6. D. B. Volkin and C. R. Middaugh, Protein solubility. In: T. J. Ahern and M. C. Manning, eds. Pharmaceuticals. Part A. Chemical and Physical Pathways of Protein Degradation. New York: Plenum Press, 1992, pp. 109–134.
7. H. R. Costantino, R. Langer, and A. Klibanov, Solid-phase aggregation of proteins under pharmaceutically relevant conditions. J. Pharm. Sci. 83(12):1662–1669 (1994).
8. G. W. Becker, et al. Isolation and characterization of a sulphoxide and a desamido derivative of biosynthetic human growth hormone. Biotechnol. Appl. Biochem. 10:326–337 (1988).
9. M. J. Pikal, et al., The effects of formulation variables on the stability of freeze-dried human growth hormone. Pharm. Res. 8(4):427–436 (1991).
10. A. Bardat, R. Schmitthaeusler, and E. Renzi, Condensable chemical vapors for sterilization of freeze dryers. J. Pharm. Sci. Technol. 50(2):83–88 1996).
11. S. E. Zale and A. M. Klibanov, Why does ribonuclease irreversibly inactivate at high temperatures? Biochemistry 25(19):5432–5443 (1986).
12. Y.-C. J. Wang and M. A. Hanson, Parenteral formulations of proteins and pep-

tides: Stability and stabilizers. J. Parenter. Sci. Technol. 42(suppl.):S3–S25 (1988).

13. D. B. Volkin and C. R. Middaugh, The effects of temperature on protein structure. In: T. J. Ahern and M. C. Manning, eds. Stability of Protein Pharmaceuticals. Part A: Chemical and Physical Pathways of Protein Degradation. New York: Plenum Press, 1992, pp. 215–248.

14. W. R. Liu, R. Langer, and A. Klibanov, Mositure-induced aggregation of lyophilized proteins in the solid state. Biotechnol. Bioeng. 37:177–184 (1991).

15. D. H. Lewis, controlled release of bioactive agents from lactide/lycolide polymer. In: M. Chasin and R. Langer, eds. Biodegradable Polymers as Drug Delivery Systems. New York: Marcel Dekker, 1990, pp. 1–41.

16. Y. Hayashi, et al., Entrapment of proteins in poly (L-lactide) microspheres using reversed micelle solvent evaporation. Pharm. Res. 11(2)337–339 (1994).

17. Y. Tabata, S. Gutta, and R. Langer, Controlled release systems for proteins using polyanhydride microspheres. Pharm. Res. 10(4):487–496 (1993).

18. S. Cohen, et al., controlled delivery systems for proteins based on poly(lactic/glycolic acid) microspheres. Pharm. Res. 8(6):713–720 (1991).

19. M. S. Hora, et al., Release of human serum albumin from poly(lactide-*co*-glycolide) microspheres. Pharm. Res. 7(11):1190–1194 (1990).

20. H. Jeffrey, S. S. Davis, and D. T. O'Hagan, The preparation and characterization of poly(lactide-*co*-glycolide) microparticles. II. The entrapment of a model protein using a (water-in-oil)-in-water emulsion solvent evaporation technique. Pharm. Res. 10(3):362–368 (1993).

21. Y. Tabata and R. Langer, Polyanhydride microspheres that display near constant release of water-soluble model drug compounds. Pharm. Res. 10(3):391–399 (1993).

22. M. Donbrow, Introduction and overview. In: M. Donbrow, ed. Microparticles and Nanoparticles in Medicine and Pharmacy. Boca Raton, FL: CRC Press, 1992, pp. 1–14.

23. Z. Zhao, et al., Controlled delivery of cytokines for therapeutic cancer vaccination. In: 21st International Symposium on Controlled Release of Bioactive Material. Seattle: Controlled Release Society, 1995.

24. E. Mathiowitz and R. Langer, Polyanhydride microspheres as drug delivery systems. In: M. Donbrow, ed. Microparticles and Nanoparticles in Medicine and Pharmacy. Boca Raton, FL: CRC Press, 1992, pp. 100–123.

25. W. R. Gombotz, M. S. Healy, and L. R. Brown, Very low temperature casting of controlled release micropsheres. U.S Patent 5,019,400 (May 28, 1991).

26. S. W. Kim. Y. H. Bae, and T. Okano, Hydrogels: Swelling, drug loading and release. Pharm. Res. 9(3):283–290 (1992).

27. J. L. West and J. A. Hubbell. Localized intravascular protein delivery from photopolymerized hydrogels. In 21st International Symposium of Controlled Release of Bioactive Msterial. Seattle: Controlled Release Society, 1995.

28. G. E. Visscher, et al., Biodegradation of and tissue reaction to 50:50 poly(DL-lactide-*co*-glycolide) microcapsules. J. Biomed. Mater. Res. 19:349–365 (1985).

29. J. M. Schakenrad, et al., Enzymatic activity toward polyL-lactic) acid implants. J. Biomed. Biomater. Res. 24:529–545 (1990).

30. C. R. Behl, et al., Optimization of sytemic nasal drug delivery with pharmaceutical excipients. Adv. Drug Delivery Rev. 29(1,2):117–133 (1997).

31. S. Calis, et al., Adsorption of calcitonin PLGA microspheres. Pharm. Res. 12(7): 1072–1076 (1995).

32. W. Norde and J. Lyklema, Thermodynamics of protein adsorption: Theory with special reference to the adsorption of human plasma albumin and bovine pancreas ribonuclease at polystyrene surfaces. J. Colloid Interface Sci. 71(2):350–366 (1979).

33. H. Thurow and K. Geisen, Stabilization of dissolved proteins against denaturation at hydrophobic interfaces. Diabetologia 27:212–218 (1984).

34. P. A. Burke, Determination of internal pH in PLGA microspheres using ^{31}P NMR spectroscopy. In: 23rd International Symposium on Controlled Release of Bioactive Materials Kyoto, Japan: Controlled Release Society, 1996.

35. K. Mader, et al., Non-invasive in vivo characterization of release processes in biodegradable polymers by low-frequency electron paramagnetic resonance spectroscopy. Biomaterial 16:1–5 (1995).

36. F. D. Anderson, et al., Tissue response to bioerodible subcutaneous drug inplants: A possible determination of drug absorption kinetics. Pharm. Res. 10(3):396 (1993).

37. M. A. Sarkar, Drug metabolism in the nasal mucosa. Pharm. Res. 9(1):1–9 (1992).

38. L. Illum, N. F. Farraj, and S. S. Davis, Chitosan as a novel nasal delivery system for peptide drugs. Pharm. Res. 11(8):1186–1189 (1994).

39. A. Yamamoto, et al., effects of various protease inhibitors on the intestinal absorption and degradation of insulin in rats. Pharm. Res. 11:1496–1500 (1994).

40. V. H. Lee, et al., Oral route of peptide and protein delivery. Peptide and Protein Delivery. New York: Marcel Dekker, 1991, pp. 691–738.

41. A. Leone-Bay, et al., 4-[4-(2-Hydroxybenzoyl)aminophenyl]butyric acid as a novel oral delivery agent for recombinant human growth hormone. J. Med. Chem. 39:2571–2578 (1996).

42. M. A. Hanson and S. K. E. Rouan, Introduction to formulation of protein pharmaceuticals. In: T. J. Ahern and M. C. Manning, eds. Stability of Protein Pharmaceuticals. Part B. In Vivo Pathways of Degradation and Strategies for Protein Stabilization. New York: Plenum Press, 1992, pp. 209–233.

43. J. F. Carpenter and J. H. Crowe, An infrared spectoscopic study of the interaction of carbohydrates with dried proteins. Biochemistry 28:3916–3922 (1989).

44. T. Arakawa, The stabilization of β-lactoglobulin by glycine and NaCl. Biopolymers 28:1397–1401 (1989).

45. K. Gekko and H. Ito, Competing solvent effects of polyols and guanidine hydrochloride on protein stability. J. Biochem. 107:572–577 (1990).

46. T. Arakawa and S. N. Timasheff, The stabilization of proteins by osmolytes. Biophs. J. 47(March):411–414 (1985).

47. T. Arakawa, Y. Kita, and J. F. Carpenter, Protein—solvent interactions in pharmaceutical formulations. Pharm. Res. 8(3):285–291 (1991).

48. S. A. Charman, K. L. Mason, and W. N. Charman, Techniques for assessing the effects of pharamaceutical excipients on the aggregation of porcine growth hormone. Pharm. Res. 10(7):954–962 (1993).

49. P. K. Tsai, et al., Formulation design of acidic fibroblast growth factor. Pharm. Res. 10(3):649–659 (1993).

50. T. A. Horbett, Adsorption of proteins and peptides at interfaces. In: T. J. Ahern and M. C. Manning, eds. Stability of Protein Pharmaceuticals. Part A. Chemical and Physical Pathways of Protein Degradation. New York: Plenum Press, 1992, pp. 195–214.

51. J. L. Cleland and A. J. S. Jones. Development of stable protein formulations for biodegradable polymers. In: Proceedings of an International Symposium on Controlled Release of Bioactive Materials. Seattle: Controlled Release Society, 1995.

52. M. Mumenthaler, C. C. Hsu, and R. Pearlman, Feasibility study on spray-drying protein pharmaceuticals: Recombinant human growth hormone and tissue-type plasminogen activator. Pharm. Res. 11(1):12–20 (1994).

53. I. Gonda, Inhalation therapy with recombinant human deoxyribonuclease I. Advanced Drug Delivery Reviews 19(1):37–46 (1996).

54. O. L. Johnson, et al., One-month release of human growth hormone from a biodegradable delivery system. In: Third U.S.–Japan Symposium on Drug Delivery Systems, Controlled Release Society, Maui, Hi, 1995.

55. O. L. Johnson, et al., A month-long effect form a single injection of microencapsulated human growth hormone. Nat. Med. 2(7):795–799 (1996).

56. M. W. Townsend, P. R. Byron, and P. P. Deluca, The effects of formulation additives on the degradation of freeze-dried ribonuclease A. Pharm. Res. 7(10): 1086–1091 (1990).

57. M. S. Hora, R. K. Rana, and F. W. Smith, Lyophilized formulations of recombinant tumor necrosis factor. Pharm. Res. 9(1):33–36 (1992).

58. K.-I. Izutsu, S. Yoshioka, and S. Kojima, Physical stability and protein stability of freeze-dried cakes during storage at elevated temperatures. Pharm. Res. 11(7): 995–999 (1994).

59. H. R. Costantino, R. Langer, and A. M. Klibanov, Mositure-induced aggregation of lyophilized insulin. Pharm. Res. 11(1):21–29 (1994).

60. S. P. Schewendeman, et al., Strategies for stabilizing tetanus toxoid toward the development of a single-dose tetanus vaccine. Dev. Biol. Stand. (in press).

61. S. Kobayashi, S. Kondo, and K. Juni, Study of delivery of salmon calcitonin in rats: Effects of protease inhibitors and absorption enhancers. Pharm. Res. 11(9): 1239–1243 (1994).

62. P. Edman and E. Bjork, Routes of delivery: Case studies (1) nasal delivery of peptide drugs. Adv. Drug Delviery Rev. 8:165–177 (1992).

63. W. J. Irwin, et al., The effects of cyclodextrins on the stability of peptides in nasal enzymic systems. Pham. Res. 11(12):1698–1703 (1994).

Index